Jacob Y. Wong
Roy L. Anderson

Non-Dispersive Infrared Gas Measurement

IFSA International Frequency Sensor Association Publishing

Jacob Y. Wong
Roy L. Anderson
Non-Dispersive Infrared Gas Measurement

E-mail (for orders and customer service enquires): ifsa.books@sensorsportal.com

Visit our Home Page on http://www.sensorsportal.com

ISBN -13: 978-84-615-9732-1
ISBN-10: 84-615-9732-X
BN-20120619-XX
BIC: TJF

About the Authors

Dr. Wong did his undergraduate work at Princeton University (AB '63) and his graduate work in infrared lasers research at Stanford University (Ph.D. '67). He joined Hewlett-Packard Company in 1967 and stayed there until 1978. After a short stay in the aerospace industry, he started a total of three companies all specializing in the R&D of advanced NDIR gas sensors. The third and latest company he started in 2003 was Airware Inc. which is still in business today. Over the years Dr. Wong has published over 40 research papers in major scientific journals and authored over 100 US and International patents worldwide.

Mr. Anderson has an undergraduate degree in chemistry from Illinois Wesleyan University (BA '79) and a Juris Doctor degree from The George Washington University (JD '82). He began with, and became a partner in, Lyon & Lyon, where he practiced patent law litigation and prosecution until 2000, when he left to practice on his own. He began working with Dr. Wong in the mid 1980's. Today he is a partner with Wagner, Anderson & Bright, P.C., he has worked on and filed dozens of patents issued to Dr. Wong relating to advanced NDIR gas sensors.

Content

Foreword

Infrared spectroscopy provides the analytical laboratory with essential capabilities to identify and to quantify components of gas mixtures in a relatively straightforward manner. Except for symmetric diatomic species, most molecules are "IR active" that is, they absorb IR light at specific energies associated with that molecule's vibrational and rotation modes. Simple molecules have a few predominant absorption energies and are easy to identify, while more complicated molecules with many bonds have many absorption peaks. To cover the full range of possible absorption energies, laboratory instruments initially employed dispersive elements, typically gratings, to scan over the wavelengths of interest. Today, Fourier-transform infrared (FTIR) spectroscopy has replaced most dispersive IR spectrometry due to improvements in speed and the signal-to-noise ratio but at the expense of instrumental complexity.

The impressive analytical power of IR spectroscopy can be distilled into a tiny sensor for a restricted, but nevertheless very useful, set of chemical vapors. Non-dispersive infrared (NDIR) sensors use bandpass filters to select one, or at most a few, energy bands corresponding absorption by carbon dioxide, water, hydrocarbons, *etc*. Although the concept is simple, the task has proved to be elusive for constructing an NDIR sensor that maintains its calibration in spite of aging and environmental factors. Over the past four decades, Dr. Wong has been on the quest to perfect NDIR sensing, yet in very practical designs. This book reflects his journey, and more recently that of his coauthor, to do just that.

Non-Dispersive Infrared (NDIR) - Gas Measurement begins with a brief survey of various gas measurement techniques. Sensor discovery and development continue to be the subject of numerous professional journals and meetings. The need for improved sensor technology in residential, commercial, automotive, military and industrial sectors has driven this interest. The majority of published papers report the observation of sensing based upon a property change of some material when it interacts with a gas (termed "Interactive" in this book). The reproducibility of such a sensor depends upon the stability of the material, which is nearly always subject to degradation over time.

Physical-based sensors, on the other hand, rely upon detecting a measurable change of some physical interaction of the gas with the sensor (termed "Non-Interactive" in this book). Most high-quality analytical instruments depend upon physical interactions. NDIR sensors fall into the category of physical based sensors, which are usually the most reliable over time, if properly engineered.

Chapter 2 gives an interesting history of infrared gas sensing and includes some well-known pioneers of spectroscopy such as Sir William Herschel and John Tyndall, but also includes the eminent Alexander Graham Bell just a few years after his invention of the telephone. One of the first practical applications of infrared gas sensing was Tyndall's quantitative measurement of carbon dioxide in the human breath (1864). Curiously a hundred years later, Dr. Wong led a team at Hewlett Packard that finally solved the issues of medical CO_2 measurement to create a practical breath monitor that served the medical community for almost two decades (Chapter 3). The development and naming of the Capnometer show how good business opportunities arise from clever science, even when initial goals are thwarted.

Dr. Wong is no stranger at the U.S. Patent Office and holds more than one hundred patents. Chapter 4 enumerates a series of NDIR patents, dominated by Wong that heralded the birth of practical CO_2 sensors and enabled Demand Control Ventilation (DCV). Because methods used in the medical Capnometer could not be afforded by DCV installations, these simpler sensors suffered calibration drift. Without Dr. Wong's scheme of Automatic Background Correction (ABC), the expense of frequent recalibration by field technicians would have sorely limited DCV applications.

Very recent design innovations promise to usher a proliferation of NDIR sensors for DCV, humidity control and safety purposes. Chapter 5 shows that issues involving the inevitable degradation and drift of components can be compensated. Thus, the intervals between calibrations could be greatly extended or even eliminated for many applications. Until now, the widespread implementation of DCV has been hindered by maintenance costs. If calibration is desired, Dr. Wong has invented a means to accomplish this using only ambient environmental air rather than bottles of calibration gas. Moreover, absolute humidity measurement has been elusive due to the difficulty in maintaining calibration. The improvements discussed in this chapter have the potential to transform the HVAC industry. Large-volume manufacture should drop costs and encourage even more widespread deployment. Very high volumes will allow microfabricated emitters and detectors to replace their less efficient predecessors. It is possible to conceive of inexpensive self-checking safety sensors for fire, methane, carbon monoxide, pollutants and toxic gases.

The reader interested in gas sensors will benefit from the following material. Much of what is known about NDIR sensing has been kept proprietary and very little has been written about its development. The authors' homey and wandering account of their developments and exploits may be amusing at times, but attest to the perseverance and good fortune required for success.

R. J. Bruce Warmack,

Senior Research Scientist, Oak Ridge National Laboratory
Oak Ridge, Tennessee,
March 2012

Chapter 1

NDIR Gas Measurement in a Nutshell

1.1. The Need for Gas Measurement

We live in a gaseous world. We breathe in air for oxygen and breathe out carbon dioxide in order to sustain the metabolism of food we eat for generating energy along with water (H_2O) we need to survive. But our respiratory system, inclusive of the trachea and the lungs, allows all manners of gas to enter our bodies without discrimination. Through bad experiences, humans early on were keenly aware of the fact that certain gases with high enough concentrations could cause ill effects inside their bodies and in some cases could even be lethal to them. But it would take the passage of many centuries before human civilization developed to the point that these circumstances could be understood. Nobody is certain when people actually possessed enough knowledge to get to the bottom of this life mystery and started to develop various methodologies for detecting different kind of gases and measuring their concentrations in the air. It is generally believed that it could not have happened within the past 500 years ! But it is also believed that once started people have not stopped continually developing newer and better methodologies for understanding this life mystery up until the present time. Furthermore, over the years people have also continued to utilize this knowledge to control the gases that they know and use them to advantage whenever or wherever an opportunity is seen.

Apart from the need to identify different kinds of gases in the air and to measure their concentration levels for health and survival reasons, people have also found out quite a bit later that the controlled combustion of certain gases in air with concentration at the right levels can be used to generate heat when the weather is cold. As much as the combustion at times was very beneficial, they also found out that it could also be very destructive if uncontrolled. People also found out that combustion of these special kinds of gases in air can also lead to deadly explosions if the concentration levels of these gases are too high. All in all, it becomes awfully clear that almost since the time when people first had the knowledge, they have been trying to determine what kind of gases are in the air that surrounds them. They have also been trying very diligently to measure the concentration level of these gases whenever and wherever they encounter them and use them to their advantage.

Over the centuries, and as far back as anyone can trace back time, people have been persistently trying to invent literally myriads of methods not only to identify gases, but also to measure their concentration levels primarily in air or in other specific gaseous backgrounds. From the gas measurement technology standpoint today, all methodologies fall under two broad classifications dependent upon their level of sophistication and areas of applications.

The first classification is called "Analytical Instrumentation," which includes gas analysis techniques such as Mass Spectrometry, Fourier Transform Infrared Spectroscopy, Gas Chromatography, Raman Spectrometry, Non-Dispersive Infrared (NDIR) and Tunable Diode Laser Absorption Spectroscopy (TDLAS). The equipment for this class of gas analysis instrumentation is generally very complex and expensive. The derived instruments are mainly used in areas of industrialization such as heavy machinery manufacturing plants, petrochemical refineries and processing plants, pharmaceutical and cosmetic factories, R&D laboratories etc. Also, the uses for such instruments would normally go far beyond simple gas measurement capabilities. As a matter of fact, in addition to gas measurement, some of them can also be used to separate and analyze chemical compounds as long as these compounds can be vaporized into gaseous form without decomposition.

The second classification is called "Sensors and Monitors" or just sensors. When compared to the first classification, the applications for sensors are totally different from those belonging to the first classification. The main functions for sensors are for safety and or security monitoring. The quantity of their usage far exceeds those for the first classification of analytic instruments. Consequently, sensors are generally rather simple, small-sized, compact and relatively inexpensive.

1.2. Interactive Gas Sensors

There are two primary categories of sensors belonging to the second classification of gas measurement technology-- "Interactive" and "Non-Interactive". In the Interactive category of sensors, a gas must physically or chemically interact with the sensor element. The gas to be sampled actually comes into contact with one or more components constituting the gas sensor (e.g. electrodes, electrolytes, sensing surfaces etc.) via oxidation, reduction, physical adsorption, etc. Inherently, this interaction over time has the effect of depleting or changing the sensor that results in a need for re-calibration and ultimately its replacement. These types of sensors, because of their interactive nature, can also be affected by other gases that may imitate or suppress the target gas reading or in some cases may completely poison the sensing element. The most common interactive gas sensors are now listed and discussed below.

1.2.1. Electrochemical Gas Sensors

Electrochemical gas sensors are by far the most important among interactive gas sensors in use today. The amperometric type of electrochemical gas detectors detects the concentration of a target gas by oxidizing or reducing it at a sensing electrode and measuring the resulting current. Another type of electrochemical gas sensors measure voltages and are called potentiometric. The amperometric sensors generally contain two electrodes in contact with an electrolyte, which is commonly a mineral acid. The sensing electrode is typically fabricated by depositing a large area of a precious metal (e.g. Platinum for detecting CO gas) onto a hydrophobic porous membrane served as a diffusion barrier which is in contact with both the electrolyte and the ambient air to be monitored. The target gas diffuses into the sensor through the back of the porous membrane to the sensing electrode where it is oxidized or reduced. This electrochemical reaction results in an electric current that passes through the external circuit. The external circuit for a two electrode sensor maintains an operating voltage between the sensing and counter electrodes to drive the electrochemical reaction in addition to measuring, amplifying and performing other signal processing functions. At the counter electrode an equal and opposite reaction occurs, such that if the sensing electrode is an oxidation, the counter electrode is a reduction.

Electrochemical sensor technology has been around for many decades and over the years has offered highly specific gas sensing in the low parts-per-million (ppm) range for many gases. While electrochemical sensors offer many advantages, particularly in size and unit cost, they are not suitable for every gas. Since the detection mechanism involves the oxidation or reduction of the gas, electrochemical sensors are usually only suitable for gases which are electrochemically active, although it is possible to detect electrochemically inert gases indirectly if the gas interacts with another species in the sensor that then produces a response. An electrochemical sensor for carbon dioxide (CO_2) which uses metal-carbonates as gas sensing electrodes is an example, although it never receives wide acceptance due to its more complicated sensor design and high unit cost.

Among the shortcomings of electrochemical gas sensors are 1) cross-sensitivity; 2) relatively short and unpredictable operating life; 3) sensor output may drift over time and 4) lack of automatic self-policing of functionality except projecting its imminent death. Despite the afore-mentioned shortcomings, the development of electrochemical CO sensors has come a long way over the past decade. Today, their sensitivity, output stability over time and operating life (minimum of 7 years) are good enough to be effectively used as CO alarms at home to detect lethal levels of CO. But more work is needed in order to further improve their detection sensitivity, output stability and even longer operating life (10-15 years minimum).

1.2.2. Solid State Electrical Conductivity Gas Sensors

This type of sensors measure the resistance change caused by the presence of a gas near a specially coated and heated surface. Because of their chemical and thermal stability, it is metal oxides that are predominantly used in this application. At temperature of approximately 500 ^{o}C the electrical resistance of many metal oxides exhibits a high degree of dependence on the concentration of certain gases in their vicinities. At temperatures of 200 – 500 ^{o}C metal oxides with n-type electron carrier paths, such as SnO_2, ZnO_2 or Fe_2O_3, primarily respond to oxidizable gases, such as H_2, CH_4, CO, or H_2S, and increase their electrical conductivity. In contrast, p-type semiconductors, such as CuO, NiO or CoO, respond primarily to reducible gases, such as O_2, NO_2 and Cl_2. The best known of this type of sensors is the Taguchi gas sensor (TGS) which is based on SnO_2. For decades Taguchi sensors have been widely used in Japan as combustible and toxic gas detectors and also installed as fire alarms in many Japanese homes.

Solid State Electrical Conductivity sensors are much less gas specific than electrochemical sensors and often lack low concentration sensitivity. Long-term drift, primarily in the form of diffusion and sintering effects at grain boundaries, is unavoidable. These sensors also tend to be more sensitive to humidity and temperature but can have a longer operating life than electrochemical sensors (4-7) years. They also have a wide gas concentration detection range but have not been applied to widespread usage outside of Japan due to their cross-sensitivity and non-specificity to the detection of different gases.

1.2.3. Photo-Ionization (PID) Gas Sensors

This type of sensors uses a process called Photo-ionization, which is applied to measuring volatile organic compounds (VOC) that can be a concern in indoor air pollution monitoring, and also in certain industrial applications. The process utilizes an ultraviolet light source to ionize target VOC molecules. The released ions are then passed by a very sensitive electric circuit that essentially counts the ions which can give an indication of gas concentrations. This type of sensors cannot distinguish different gases but can identify all gases present that are over a target ionization level, which in many cases are useful for indicating relative toxicity. The Photo-Ionization (PID) sensors are non-destructive detectors. They do not appreciably destroy or consume the components they detect. However, the signal produced by a PID sensor may be quenched when measuring in high humidity environments, or when a compound such as methane (CH_4) is present in high concentration. This attenuation is due to the ability of water, methane, and other compounds with high ionization potential values to absorb the photons emitted by the ultraviolet lamp without leading to the production of ion current. In other words, this will reduce the number of energetic photons available to ionize the target molecules. Typical PID sensors measure VOCs and other gases in concentrations from sub parts per billion (ppb) to 10,000 parts-per-million (ppm).

1.2.4. Flame-Ionization (FID) Gas Sensors

As the name suggests, analysis by this type of sensors involves the detection of ions. They are therefore also akin in operating principles to

Photo-Ionization (PID) sensors. However, for Flame-Ionization (FID) sensors, the source of the ions is a small hydrogen-air flame, therefore an FID sensor destroys most, if not all, of the components it is detecting. Contrarily, for PID sensors, the ions are from a process called photo-ionization generated with an ultraviolet light source, the components can continue on to another detector after passing through the PID sensor. With an FID sensor, most components are destroyed and no further detection is possible.

The detection of organic compounds is most effectively done with FID sensors, particularly biochemical compounds such as proteins, nucleotides, and pharmaceuticals, which have a greater amount of carbon present than other elements. This means that a particular compound may be more easily detected using FID sensors over other sensors because of higher carbon concentration and also FID's outstanding sensitivity in detecting it. Generally speaking, FID sensors are best for detecting hydrocarbons and other easily flammable components. They are very sensitive to these components, and response tends to be linear across a wide range of concentrations.

1.2.5. Colorimetric Gas Sensors

This type of sensors incorporate a chemical coating that will change color in proportion to the amount of gas the sensor is exposed to. The color change is cumulative over time. As a result, many sensors like this are used as simple warning indicators of the presence of the gas if a rapid color change is observed. Some gas sensors use special light reading elements to note the progress of the color change to determine the current gas concentration. These sensors currently are used in a very small percentage of gas sensing applications.

1.2.6. Catalytic Bead Gas Sensors

This type of sensors is used principally for the measurement of combustible gases. The sensor consists of two coils of fine platinum wire each embedded in a bead of alumina, connected electrically in a Wheatstone bridge circuit. One of the pellistors is impregnated with a special catalyst which promotes oxidation while the other is treated to inhibit oxidation. Current is passed through the coils so that they reach a temperature at which oxidation of a gas readily occurs at the catalyzed bead (550–550 °C). Passing combustible gas raises the temperature further which increases the resistance of the platinum coil in the catalyzed bead, leading to an imbalance of the bridge. This output change is linear for most combustible gases, up to and beyond 100 % LEL, response time is a few seconds to detector alarm levels (around 20 % LEL), but at least 12 % oxygen by volume is needed for the oxidation. Typically these sensors have a 3-year life and require regular calibration particularly in critical life/safety applications related to petrochemical processing and handling.

1.3. Non-Interactive Gas Sensors

Unlike the interactive category of gas sensors discussed above, when a gas to be sampled comes into contact with one or more components constituting the gas sensor (e.g. electrodes, electrolytes, sensing surfaces etc.) via oxidation, reduction, physical adsorption etc., detection of a gas by a non-interactive gas sensor involves none of the above. The non-interactivity refers only to the state of the gas before and after its measurement by the sensor. Simply put, a non-interactive gas sensor does not and will not change in a significant way the state of the gas it measures.

Just about all of the non-interactive gas sensors in use today are optical gas sensors. By optical what is meant is that only electromagnetic radiation or light of different wavelengths comes into contact with the sampled gas to make the necessary measurement. The radiation does interact with the gas and a portion of it may even be absorbed by it in order to change the excitation state of the gas to be sampled. However, the excited gas molecules subsequently return to their ground states quickly via collisions with the sample chamber walls or with other gas molecules and in the end nothing has really been changed, either physically or chemically, to the sampled gas itself in the measurement process. The wavelength of electromagnetic radiation used in non-interactive gas sensors can span from the ultraviolet (< 3,000 Å) all the way to the far infrared (> 15.0 μ). The electromagnetic radiation can either be coherent or incoherent. The most common non-interactive gas sensors are now listed and discussed below.

1.3.1. Tunable Diode Laser Absorption Spectroscopy (TDLAS) Gas Sensors

TDLAS is a superior gas measurement technique, which became fully developed only within the past three decades, for measuring the concentration of gas species in a gaseous mixture using tunable diode lasers and laser absorption spectrometric techniques. A basic TDLAS gas sensor consists of a tunable diode laser light source, transmitting optics, sample chamber containing the gas to be sampled, receiving optics and appropriate detectors dependent upon the wavelength of the diode laser used. The emission wavelength of the tunable diode laser is tuned over a selected absorption line of a gas in the path of the laser beam. The characteristics of the selected absorption line must be spectrally narrow and strength-wise strong enough such that a reduction of the measured signal intensity, which can be detected by a photodiode, can be used to determine the gas concentration and other properties of the gas such as its temperature and pressure.

Different diode lasers are used based upon the application and the range over which tuning is to be performed. Typical examples are InGaAsP/InP (tunable over 900 nm to 1.6 µ), InGaAsP/InAsP (tunable over 1.6 µ to 2.2 µ), etc. TDLAS gas sensors did not emerge in significant numbers until after the early 1990's. But as early as the late 1980's, one of the present authors was among the first researchers to propose using this technique for the concentration measurement of gaseous oxygen utilizing one of the sharp lines of its "A" band centered at 7,600Å and an GaAlAs diode laser (Wong, 1988a). Despite the fact that TDLAS gas sensors are today widely used for quantitative assessments of species in the gas phase, the absorption spectrometric technique they depend upon has its own limitations. Technically speaking, the main disadvantage of this technique is that it relies on a measurement of a small change of a signal on top of a large background. Any noise introduced by the light source or the optical system will deteriorate the detection capability of the technique. The sensitivity of direct laser absorption techniques is therefore limited to an absorbance of $\sim 10^{-3}$, insufficient for many types of applications. Another limitation of this technique is the availability and high cost for the diode lasers needed in making the sensors.

1.3.2. Non-Dispersive Infrared: Photo-Acoustic and Optical Gas Sensors

TDLAS and Non-Dispersive Infrared (NDIR) gas sensors both belong to the non-interactive category of gas sensors. The former uses coherent radiation and the latter uses incoherent radiation as their respective light sources to make gas measurements. There are two types of non-interactive NDIR gas sensors, namely NDIR Photo-Acoustic and NDIR Optical. The former type is also known as Photoacoustic Spectroscopy (PAS) and the latter simply known as NDIR. Interestingly enough, these two gas measurement techniques are based upon the same Physics principle, namely, the ability of gases to absorb infrared light. However, the similarity ends right there. We shall first discuss briefly the PAS gas sensors followed by the NDIR gas sensors.

For PAS gas sensors, the gas to be measured is irradiated by modulated light of a pre-selected wavelength. The gas molecules absorb some of the light energy and convert it into an acoustic signal which is detected by a microphone. The IR source is simply a heated blackbody. A mirror focuses the light onto the window of the PAS cell (sample chamber) after it has passed the light chopper and the optical filter. The chopper is basically a slotted disk that rotates and effectively “switches” the light on and off. The optical filter is a narrow-bandpass IR interference filter.

After passing through the window, the light beam enters the PAS cell. If the frequency of the light coincides with an absorption band of the gas in the cell, the gas molecules will absorb part of the light. The higher the concentration of the gas in the cell, the more light will be absorbed. As the gas absorbs energy, it is heated and therefore expands and causes a pressure rise. As the light is frequency chopped, the pressure will alternately increase and decrease – an acoustic signal is thereby generated. For better detection sensitivity, the acoustic signal is typically detected by two microphones, one on each side of the PAS cell. The electrical output signals from the two microphones are added in an amplifier before they are processed. But this type of measurement is highly sensitive to ambient vibrations and minor fluctuations in pressure. As a result, most attempts to make a low cost, stable gas measurement platform utilizing PAS gas sensors have been difficult because of the range of vibrations and pressures found in the real world.

It is important to note that the development of PAS gas sensors started during the first half of the twentieth century before the invention of many electronic components, devices and integrated circuits, including diodes, transistors and the like. That is the reason why the interpretation, utility and application of Physics principles to everyday life use had to be, during that time, expressed and channeled only through the understanding of traditional Physics and Chemistry. It is therefore not surprising at all to see that the development and subsequent success of NDIR gas sensors had to follow PAS gas sensors as the electronic industry finally came to life starting from the second half of the twentieth century.

As mentioned earlier, both the PAS and NDIR gas sensors share the same operating principle, namely the Physics of infrared spectroscopy. Simply explained, all gas molecules, due to their unique individual molecular makeup, vibrate at a particular identifiable frequency. This vibration/rotational frequency causes the molecules to absorb light at very specific wavelengths in the infrared spectrum. Targeted gas measurement can therefore be made by measuring the absorption of infrared light at its corresponding wavelength. Whereas the PAS technique measures this infrared light absorption as a pressure increase signal via a very sensitive microphone, the NDIR technique measures directly the same via the output of a very sensitive infrared detector, such as a thermopile. As can be seen a wide variety of gases can be measured using this infrared absorption approach, depending on the wavelength the sensor is tuned to.

Starting almost over two decades ago, gas measurement using the PAS technique has slowly been in the decline in favor of the NDIR technique mainly because of the availability of advanced electronic components and devices, particularly the microprocessor. Since then only NDIR gas sensors prevailed and are now considered as one of the best methods of gas measurement since the 1990's. Today they are reasonably priced and widely employed in many applications. They are also highly gas specific, sensitive, fast responding, relatively stable over time, rugged, reliable and easy to operate and maintain.

1.3.3. NDIR Gas Sensors in a Nutshell

To put gas measurement technology for NDIR gas sensors in a nutshell, let us first briefly review its operating principle for gas detection. NDIR gas sensors use the infrared spectrum between 2 and 20 μ in gas detection because it is here that many gases show characteristic vibration/rotation absorption spectra with narrow, non-overlapping bands. Figure 1 shows the infrared signatures for a number of common and important gases in the infrared spectrum. Note that in Figure 1 most of the gases possess their own unique infrared signatures so that it is possible to make an NDIR gas sensor which is selectively sensitive to a specific gas or group of gases.

After understanding the operating principle for NDIR gas sensors, let us now elucidate how it is actually being applied to gas detection. One starts off with the four components constituting the basic sensor as portrayed in Figure 2. With reference to Figure 2, the four components are the source, the waveguide, the infrared filter and the infrared detector.

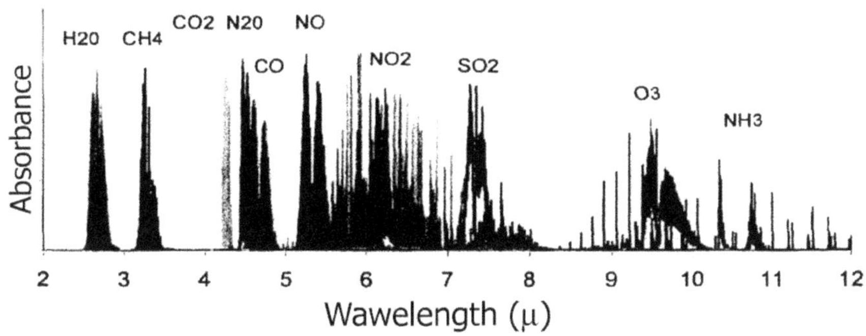

Figure 1. Absorption Signatures of Various Gases in the Infrared Spectrum.

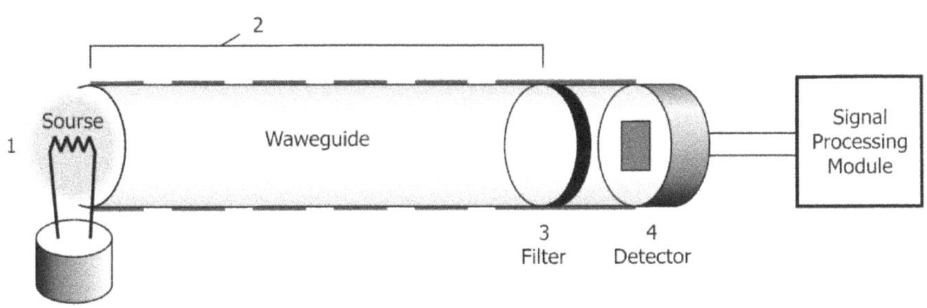

Figure 2. The Four Basic Components of an NDIR Gas Sensor.

1.3.3.1. The Source

The source 1 provides the infrared energy for the gas measurement. As it turns out over the years, the most reliable source is actually a small incandescent lamp (miniature light bulb) that gives off a broad band of energy that includes infrared energy. The only drawback to this type of source is that the glass envelope of the light bulb blocks most infrared energy beyond 4 microns in wavelength, which means that many gases in the infrared spectrum (see Figure 1) cannot be measured. There are other non-bulb sources that can emit longer wavelength radiation but these types of sources tend to be a lot more expensive and their spectral output may not be as stable as the light bulb over time. Bulb sources also have problems with the infrared output changing over time. As a matter of fact, for decades source output instability over time has been plaguing the infrared gas sensor industry resulting in sensor output drifts and the need for periodic ongoing recalibration. One advantage to these other types of sources is that their power consumption can be much less that the light bulb, because they just emit the needed infrared energy in the spectral band of interest rather than the bulb which is a quasi-blackbody source that radiates a broad band of energy in the infrared spectrum. Alternative sources include infrared emitting diodes (IREDs), which are very expensive but are starting to appear and become available, and MEMS sources (Micro-Electro-Mechanical-System devices). Today the light bulb is still the most cost effective and the most reliable infrared source available to NDIR gas sensor manufacturers for making sensors that detect target gases with signatures shorter than 4 μ. It is also important to note that in order to discriminate against copious but unwanted ambient infrared radiation coming everywhere from a sensor's surroundings, the signal processor of NDIR gas sensors must operate in an AC mode. Fortunately this can easily be accomplished by simply pulsing the miniature light bulb at a relatively low frequency (1-3 Hz typical) for its use.

1.3.3.2. The Waveguide

With reference to Figure 2, the waveguide 2 is the space where gas is introduced into the sensor. It places the gas in the direct path of radiation from the source that moves through the waveguide, then onto the infrared detector at the opposite end of the device. It is called a waveguide because the internal surface of the guide is usually highly reflective to infrared radiation and serves to focus light through the gas onto the detector without the need for special optics, which can add significant costs to the sensor. The gas to be measured can either be filtered or pumped into the waveguide, or a diffusion membrane can be used. The diffusion membrane is a passive filter device that allows gas molecules to move through the filter but prevents much larger particles from entering into the sensor. By filtering out particles, a major potential source of drift of the sensor can be eliminated. This is because particles inside the sensor can change the optics of the system (waveguide reflectivity) over time causing sensor drift. It is also important to note that the movement of gas into the chamber is based upon the principle of diffusion--that is the tendency for gas molecules to equalize their spacing in air to provide a uniform concentration. To the gas molecules, the diffusion membrane is like open air so that molecules can easily move in and out of the waveguide based on the concentration of the gases in the environ the sensor is in.

1.3.3.3. The Infrared Filter

With reference to Figure 2, the infrared filter 3 is the non-dispersive element designed to only let light through, coming from the infrared source after traversing the waveguide and the sample of gas that is to be measured, that corresponds to the absorption band of the gas of interest. The more molecules there are in the waveguide blocking light at this key wavelength, the less light gets through the filter to the detector. The reduction of light at the particular wavelength can be correlated to the number of gas molecules in the waveguide and consequently the gas concentration in the air. It is interesting to note that the name of this gas measurement technique, namely Non-Dispersive Infrared (NDIR), came about because of the use of the infrared filter. This multilayer thin film filter in the sensor design is considered as a non-dispersive element since it, unlike a prism or a grating, does not separate wavelengths spatially. Gratings are found to be significantly more efficient in terms of spectral transmission than prisms (3 to 100 times better depending on the dispersion of the prism material). Optical filters are even more efficient than gratings (typically more than 1,000 times better). NDIR gas sensors using optical filters therefore have a clear performance advantage over sensors designed using dispersive elements. Consequently the design of gas sensors utilizing infrared absorption has always preferentially used NDIR techniques.

1.3.3.4. The Infrared Detector

The amount of light transmitted through the infrared filter impacts on the infrared detector 4 (see Figure 2). This device is similar to many of the detectors used in motion sensors for security purposes. The net increase or decrease in light to the detector is translated into the gas concentration measured. For reasons of simplicity and clarity we shall just skim through this particular component for NDIR gas sensors. In reality, the topic of infrared detectors itself is enormously wide and complicated and one can easily write a book or two about them. As far as NDIR gas sensors are concerned, suffice it to say that their ultimate performance and unit cost in measuring gases in general are a very strong function of what kind of detectors are used in their design. As an example, one can design an NDIR gas sensor for a particular gas with a resolution of say ±10 ppm with a certain detector costing say, $ 10 each. If one should be allowed to use another detector that is costing $ 1,000 each but with noise performance (or its D_λ^* value) 100 times better, it becomes possible that with this detector one can design an NDIR gas sensor for measuring the same gas but with a resolution better than ±100 ppb or 100 times better.

1.3.3.5. Putting the NDIR Sensor Components Together

After understanding how the components or building blocks for an NDIR gas sensor are put together (see Figure 2), along with their respective performance functions, we are ready to translate their actions or outputs quantitatively and express them in such a way in order to show how such a sensor not only can identify the target gas, but also is able to measure its concentration level. Before we proceed, let us briefly review, step by step, how infrared detection is being carried out inside an NDIR gas sensor as illustrated in Figure 3. With reference to Figure 3, Figure 3(a) shows the infrared gas sensor and its key components. Figure 3(b) shows the situation when there are only a few of the gas molecules of interest found inside the waveguide; in this case, the sensor may not detect such a low level. Figure 3(c) shows the situation when more of the target molecules enter the waveguide and absorb infrared energy at the target wavelength, the amount of infrared energy or light reaching the detector through the infrared filter is decreased. Figure 3(d) shows the situation when increasingly more molecules enter the waveguide, the infrared energy measured by the infrared detector would further decrease. Figure 3(e) shows the situation when other types of gases that have different molecular makeups and therefore different infrared signatures enter the waveguide, there would be no effect on the sensor reading.

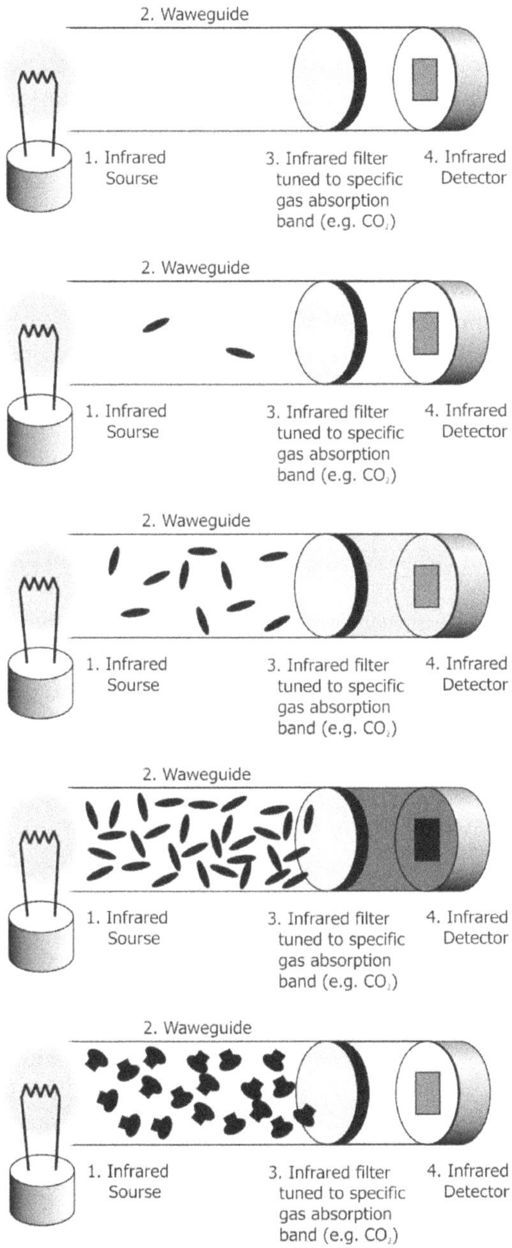

(a) Key components.

(b) Just a few molecules, no discernable detector output.

(c) More molecules, more energy absorbed, detector output decreases.

(d) Even more molecules, even more energy absorbed, detector output further decreases.

(e) Other irrelevant molecules, no matter how many, have no impact on detector output.

Figure 3. Infrared Detection Step-by-Step.

1.3.4. The Technical Foundation of NDIR Gas Sensors

In the above sections we have qualitatively described for NDIR gas sensors their operating principle, their design construct and the sensor components with their respective functions. We are now in a position to forge them together with additional technical information into an actual working gas sensor. The technical foundation of NDIR gas sensors is based upon the law of absorption known as the Beer-Lambert Law:

$$I/I_0 = \exp(-kCL), \qquad (1)$$

where:

I_0	=	Initial radiation beam intensity;
I	=	Beam intensity after traversing the gas to be detected;
k	=	Absorption coefficient = $AC/(0.434 \times C \times L)$;
AC	=	Absorbance;
C	=	Gas concentration;
L	=	Sample path length defined typically by the effective sample chamber length of the sensor;
I/I_0	=	Transmission = 1 – Absorption = $\exp(-AC/0.434)$.

One can see from Equation (1) above that the absorption of a beam of radiation traversing a sample of gas whose concentration level is to be determined is proportional to the absorption coefficient k of the gas in question, the gas concentration C and the path length L as defined by the sample chamber length for the sensor. Alternatively, the same absorption can be related directly to the value of the Absorbance (AC) of the gas in question as illustrated in Figure 4 for the CO_2 gas.

Figure 4 portrays the 4.26 μ absorption band of CO_2 gas plotting the Absorbance value of the gas versus wavelength for a concentration of 100 ppm in N_2 and a path length of 1 meter in 1 atm. pressure and a temperature of 25 °C. In general, the strength for the absorption band for a gas is reflected directly by its Absorbance value. The higher the Absorbance value of a gas, the stronger will be its absorption band. As can be seen from Figure 4, the 4.26 μ absorption band of CO_2 comprises actually a number of vibration-rotational narrow absorption lines constituting the P and the R branches. Radiation emanating from the infrared source whose spectral positions (wavelengths) coincide with the narrow absorption lines of the P and R braches will be absorbed by the CO_2 gas as the radiation traverses the sample chamber containing same. The amount of infrared energy absorbed by the gas of interest present in the waveguide which can be translated indirectly into its concentration level can be determined quantitatively with the use of Equation (1) or the Beer-Lambert absorption law stated above.

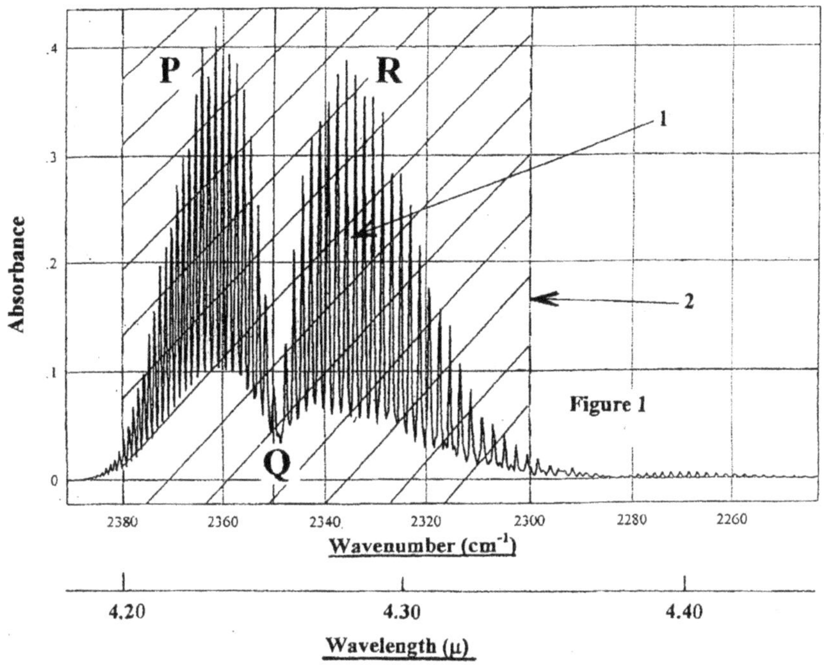

Figure 4. CO_2 Absorbance vs. Wavelength.

At this point we have just about covered all the basics with regards to an NDIR gas sensor. Enough information has been provided to the reader such that he or she can now proceed to put together and fabricate such a device. However, what is still missing, probably the most important part of this undertaking, is the design of the sensor itself. The information about NDIR gas sensors presented so far above has never been gathered together in one place like what is being put together here, at least in the opinion of the present authors. NDIR gas sensors have been around for more than five decades and over the years dozens of manufacturers for these sensors have come, stayed and gone. One can conjecture about the reason why this is so. In the opinion of the present authors, it is just a logical outcome expected of a highly competitive business environment. During the past thirty some years, no one in the industry was willing to share any technical knowledge or privileged information publicly about how to design and build NDIR gas sensors. As a matter of fact, even one of the present authors has constantly been offered opportunities to write a book about NDIR gas sensors during this period and had always politely reclined to do so. So the fact of the matter is that even if one is provided with all the necessary information about an NDIR gas sensor, it is not sufficient to come up with a design that will work well or even stand out from the others. One of the present authors has spent over fifty years in this field, and in his opinion, only one thing stands out to be true about the design of an NDIR gas sensor. And the truth is that the best design is always driven only by the performance goal set out for it right from the beginning.

Put simply, the design of an NDIR gas sensor by anyone is and should be governed only by the performance objectives expected of it. The range of desirable sensor performance objectives is indeed very wide. Needless to say, the expectation has been getting even higher and higher over the years but the basics still remain relatively unchanged. The gas sensor's principal performance objectives have always been hovering around 1) Detection sensitivity and resolution; 2) Detection range; 3) Available output modes including relays; 4) Response time; 5) Output stability over time; 6) Operating life expectancy; 7) Interference rejection capability; 8) Ruggedness; 9) Sensor dimensions or size; 10) No moving parts and, last but not least, 11) Unit cost. Generally speaking, depending upon a sensor's most important performance objectives, it can be designed accordingly.

Against the NDIR gas sensor design background presented above, there are four basic sensor design configurations developed, understood and evaluated over the years that are worthy of presentation and discussion. They are 1) Single Beam; 2) Double Beam; 3) Negative-Filtering Double Beam and 4) Absorption Biased Double Beam. Each of these sensor design configurations will be briefly discussed below. It is important to note that the four basic sensor design configurations chosen to be described here represent in no way the totality of NDIR gas sensor design methodologies. The following methodologies are presented strictly because of the opinion of the authors and also of their limited knowledge in this field. Readers are encouraged to explore on their own elsewhere additional NDIR gas sensor design methodologies whenever and wherever they are available.

1.3.4.1. Single Beam Configuration Design

Figure 5 depicts schematically the configuration layout for the Single Beam NDIR gas sensor design. With reference to Figure 5, radiation emanating from the infrared source S traverses the waveguide sample chamber SC wherein the gas G to be measured resides and impinges upon the detector D after passing through the spectral filter F located immediately in front of it. This arrangement works as an infrared gas sensor as long as the pass band of the filter F coincides with the absorption band of the gas in question. This Single Beam sensor design represented one of the earliest designs for NDIR gas sensors prevalent in use during the late 1950s to the early 1960s. The only difference is that the waveguide sample chamber SC of Figure 5 replaces the more conventional refractive and/or reflective optics used during that era. One problem associated with the Single Beam design described above is that factors other than gas can lead to a change in the detected signal, causing a measurement error. These factors include changes in the source output, the filter spectral transmission and detector responsivity with both time and temperature, not to mention errors that might be caused by the signal processing electronics and the fouling of the sensor optics. Despite these inherent shortcomings of the Single Beam design, some NDIR gas sensor manufacturers today still adhere to this design for their products primarily for reasons of sensor cost. We will have occasion to return to this seemingly illogical point and attempt to come up with some additional reasons to explain it later on in this book.

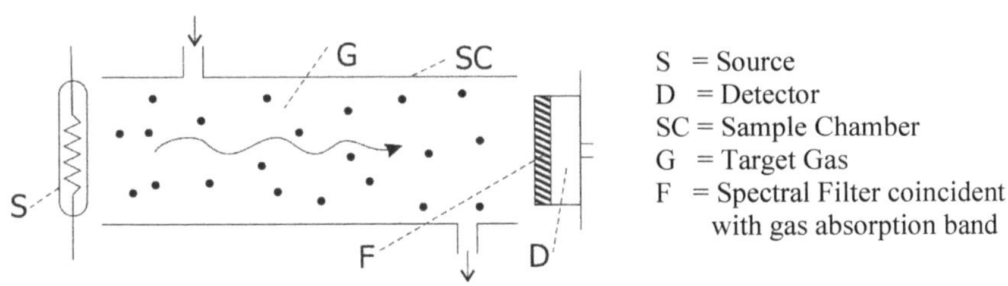

Figure 5. Configuration Layout for Single Beam.

1.3.4.2. Double Beam Configuration Design

An effective way of eliminating or reducing the error-causing factors inherent with the Single Beam sensor design mentioned above is to include a second channel which operates at a neutral wavelength just off the absorption one but in no way coincident with it. This so-called reference channel is equally affected by all non-gas phenomena as the signal channel, so by taking the ratio of the signal outputs from the two channels, namely the signal and the reference, one can ideally eliminate in theory the effect of all the error-causing factors inevitably present in the Single Beam. Only the presence of the target gas in the optical path will affect the ratio. Figure 6 depicts schematically the configuration layout for this so-called Double Beam NDIR gas sensor. With reference to Figure 6, radiation emanating from the source S traverses the waveguide sample chamber SC wherein the to-be-measured gas G resides, and impinges just about equally on both detectors D1 and D2 carrying respective filters F1 and F2. D1 is the signal channel detector carrying spectral filter F1 whose center wavelength (CWL) is coincident with the gas absorption band. D2 is the reference channel detector carrying filter F2 whose center wavelength (CWL) lies outside of the gas absorption band. The ratio of the detector signal outputs for the signal and the reference channels is processed as the output for the sensor.

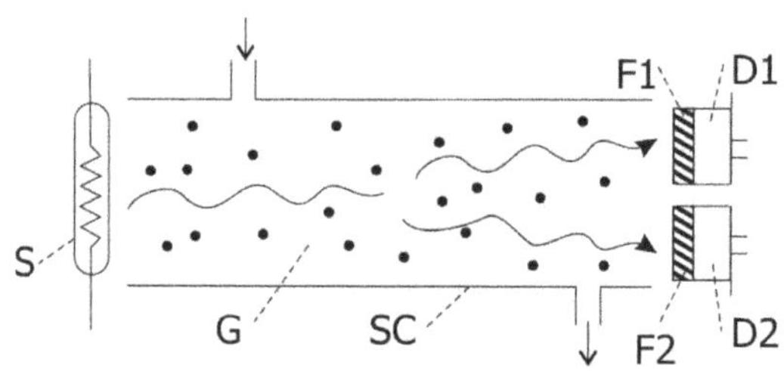

S = Source
D1 = Digital Channel Detector
D2 = Reference Channel Detector
SC = Sample Chamber
G = Target Gas
F1 = Signal Channel Spectral Filter coincident with gas absorption band
F2 = Reference Channel Spectral Filter outside of gas absorption band

Figure 6. Configuration Layout for Double Beam.

Double Beam designed NDIR gas sensors have been in use since the early 1960s after the declassification of the narrow band pass filter for public use by the US Defense Department in the late 1940s. Today they are still the mainstay NDIR gas sensors manufactured and in use. Double beam designed NDIR gas sensors are inherently more stable than their Single Beam counterparts and are preferred for applications where system integrity is a prime consideration, such as in the detection of hydrocarbons in petrochemical plants. On the other hand,

CO_2 sensors used for monitoring human presence in building ventilation systems in order to save energy using Demand Control Ventilation (DCV) strategy tend to use the simpler and less expensive Single Beam approach.

1.3.4.3. Double Beam Saturation-Filtering Design

While the Double Beam designed NDIR gas sensors are in many performance respects better than their Single Beam counterparts, there remain a number of performance deficiencies that still need to be addressed today. By far the most important one is sensor output instability over time. Another one is output dependence as a function of the temperature of the surroundings in which the sensors are located. According to the rationale for the design of the Double Beam methodology, the addition of a reference channel operating at a different wavelength off the signal channel absorption one and then processing the ratio of the signals of the two channels as the sensor output will eliminate or reduce many error-causing factors inherent with the Single Beam design. The reason is that since both channels are equally affected by non-gas factors, processing their ratio will offset the ill effects of them. These non-gas factors include changes in the source output, the filter transmission, detector responsivity, signal processing electronics and fouling of the optics over both temperature and time. While it is generally true that the Double Beam design will eliminate or reduce most of the non-gas factors if they significantly affect both the signal and the reference channels, Double Beam designed NDIR gas sensors still suffer from output drifts over time and environmental temperature dependences. Thus, the addition of an off-absorption wavelength reference channel is clearly not the cure-all solution.

As it turns out, if the output of the source changes over time, not only in magnitude (radiation intensity) but also in spectral content, then both the reference and the signal beams will be affected differently by this non-gas factor. In the design for NDIR gas sensors, almost all infrared sources without exception are graybodies, so-called because the emissivity of these sources is always less than unity. However, Planck's radiation law comprising a family of spectral radiation curves (called spectral radiant excitance curves) for sources of diverse temperatures represents equally well both graybody (emissivity < 1) and blackbody (emissivity $= 1$) sources. Figure 7 shows the spectral radiant excitance curves of a blackbody source at temperatures between 100 and 1,000 ^{o}K. As an example in the current discussion, for an

NDIR CO_2 gas sensor designed according to the Double Beam methodology presented earlier, the absorption wavelength chosen for the signal channel is CWL = 4.26 μ and the off-absorption wavelength chosen for the reference channel is CWL = 3.91 μ. The respective spectral locations of these two wavelengths are shown superimposed on the Planck curves in Figure 7 as vertical straight lines 3 and 4. One can see from Figure 7 that if the temperature of the source changes, thereby shifting its Planck's spectral radiant excitance curve, the relative magnitude of the radiation traversing the two filters at their respective spectral locations will change and so will the ratio of their respective detector outputs. This is precisely the reason why the outputs of all NDIR gas sensors today that utilize this so-called "Double Beam" methodology drift over time as the infrared source ages, resulting in a change of its spectral output intensities at the signal and reference filter wavelengths.

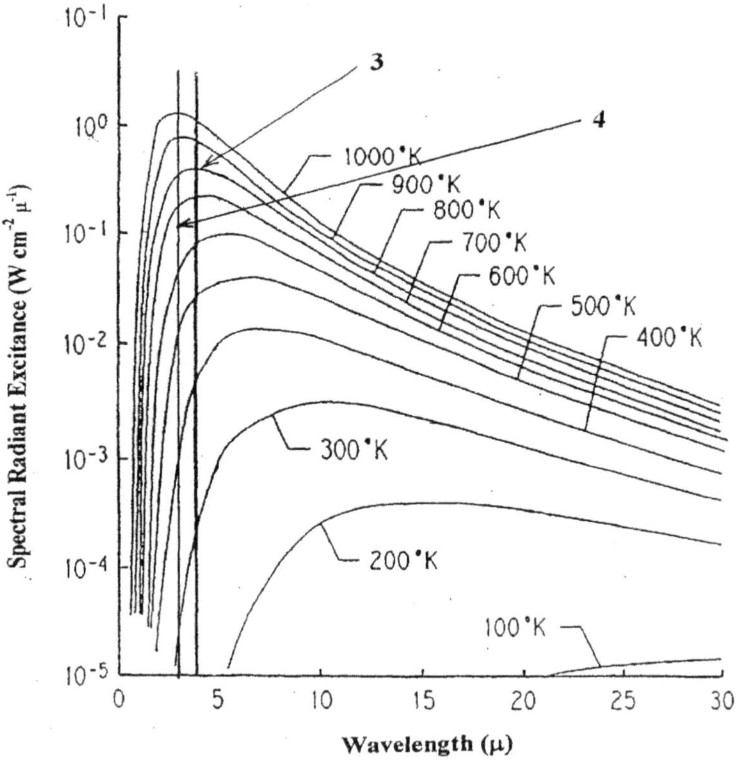

Figure 7. Spectral Radiant Excitance for a Blackbody at Temperatures 100 – 1000 °K.

For sensors using an incandescent source like a miniature light bulb, the use of the Double Beam approach is further compromised because a light bulb is actually quite a complex graybody radiation source. It contains a very small filament that operates at around 700 °C and a large surface area glass bulb envelope that is at around 100 °C. This structure tends to make the actual blackbody temperature measured in the system difficult to predict because the envelope temperature, which is affected by the temperature of the surroundings in which the sensor is located, in turns affects the temperature of the bulb filament which is the main source of the sensor radiation. Thus, the environmental temperature, a non-gas factor which supposedly does not affect the sensor output for a Double Beam design, turns out not to be the case for sensors using an incandescent lamp radiation source.

From the discussion above, it becomes very clear that in order to design an NDIR gas sensor whose output remains substantially drift free over the course of time, one must make sure that there is no spectral content discrepancies delivered at any time from the source to both the signal and the reference channel detectors. Only in this way will the ratio of the signal detector output over the reference detector output stay substantially invariant even as the infrared source ages or as the temperature of the environment surrounding the sensor changes unpredictably over time. The Double Beam Saturation Filtering methodology, disclosed less than a year ago (Wong, 2011a), accomplished just this feat. This methodology utilizes a single (spectrally identical) narrow band-pass filter in a Double Beam NDIR gas sensor for both the Signal and the Reference channels wherein the Reference channel also contains a saturation cell containing the gas of interest with both the signal and the reference detectors mounted on a common thermal platform.

Figure 8 depicts the optical component layout for the Double Beam Saturation Filtering NDIR gas sensor. As shown in Figure 8, both the Signal channel 9 and the Reference channel detector 10 are entrapped with 100 % dry Nitrogen 11 and have the same narrow band-pass spectral filter 12 which is used to detect the gas of interest in the sample chamber 13. As an example, the filter designed to be used for the detection of CO_2 gas could have a Center Wavelength (CWT) = 4.26 μ and a full Width Half Maximum (FWHM) = 0.14 μ. Both detectors 9 and 10 are thermally connected to the sensor sample chamber 13 which is in essence an aluminum waveguide sample chamber. Consequently, detectors 9 and 10 always share the same thermal platform with each other. One end 14 of the sensor sample chamber 13 is the source mount 15 housing the common infrared source 16. The other end 17 of the sensor sample chamber 13 is partitioned out respectively into the Signal channel portion 18 and the Reference channel portion 19. The signal channel detector 9 is located at the end of the Signal channel portion 18. The Reference channel detector 10 is located at the end of the Reference channel portion 19 after the saturation cell 8. Because of the fact that detectors 9 and 10 share the same common thermal platform, no significant temperature difference exists at any time between the two detectors that will alter in any way their respective responsivities or spectral responses. Thus the ratio of the signal channel voltage output over the Reference channel voltage output will virtually be independent of the ambient temperature of the environment in which the sensor is located.

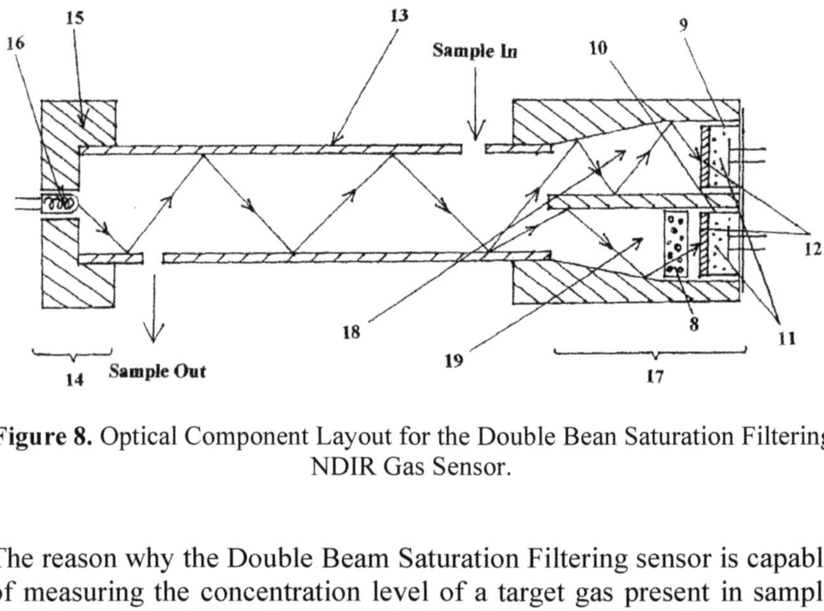

Figure 8. Optical Component Layout for the Double Bean Saturation Filtering NDIR Gas Sensor.

The reason why the Double Beam Saturation Filtering sensor is capable of measuring the concentration level of a target gas present in sample chamber 13 is due to the strategic installation of saturation cell 8 in front of the Reference channel detector 10 (see Figure 8). The function of saturation cell 8, which contains a concentration c' of the gas of interest, is to create an absorption bias between the Signal and the Reference channels. Because of saturation cell 8 and its additional absorption of source radiation, the output signal of the Reference channel detector 10 is always less than that of the Signal channel detector irrespective of the amount of the target gas present in the sensor sample chamber 13. When there is no target gas present in the sensor sample chamber 13, the ratio of the Signal channel detector output over that for the Reference channel detector remains always substantially unchanged. This is because of the fact that both channels are equipped with the same spectral narrow band-pass filter and the ratio of their detector outputs is therefore not affected by any output change, spectrally or intensity-wise, of the common infrared source. The absorption bias between the Signal and the Reference channels created by the saturation cell 8 enables the sensor output to be calibrated against the concentration level of the target gas present in the sensor sample chamber 13.

1.3.4.4. Double Beam Absorption Biased Design

Another NDIR gas sensor design capable of overcoming the sensor output drift problem over time suffered by the convention Double Beam technique is the Double Beam Absorption Biased methodology disclosed recently (Wong, 2011b). In this methodology an absorption bias between the Signal channel output and the Reference channel output for a conventional Double Beam NDIR gas sensor is created by using a signal channel with a path length that is greater than that of the reference channel. Both the Signal and Reference channel detectors are equipped with an identical narrow band-pass filter with the same Center Wavelength (CWL), Full Width Half Maximum (FWHM) and transmittance efficiency (η) at the CWL. The Signal and the Reference channel detectors also share the same thermal platform at all times.

Figure 9 depicts the optical component layout for the Double Beam Absorption Biased NDIR gas sensor. As shown in Figure 9, both the Signal channel detector 5 and the Reference channel detector 6 are entrapped with 100 % dry Nitrogen 7 and are equipped with the same narrow band-pass spectral filter 8 which is used to detect the gas of interest present in sample chamber 9. As an example, the filter designed to be used for the detection of CO_2 gas could have a Center Wavelength (CWL) = 4.26 μ and a Full Width Half Maximum (FWHM) = 0.14 μ. As shown in Figure 9, both detectors 5 and 6 are thermally connected to the entire sensor body 10 through their respective waveguides 11 and 12 and consequently they always share the same thermal platform with each other. In other words, the entire sensor body 10, which is in essence a composite of aluminum parts comprising the infrared source mount 13, sample chamber 9 and the waveguides 11 and 12, respectively, for the Signal and Reference channels, provides an excellent common thermal platform for detectors 5 and 6. Because of the fact that detectors 5 and 6 share the same common thermal platform, there should be no difference in temperature between the two detectors, or at least not any difference significant enough to interfere with their measurements over time since the thermal platform serves to equilibrate over time any localized temperature gradient that exists.

As shown in Figure 9, the sample chamber pathlength L_R, 14, associated with the Reference channel is approximately one-half of the sample chamber pathlength L_S, 15 associated with the Signal channel. Although this difference in pathlength is an important design parameter for the sensor, the difference in pathlength can be greater or less, as long as a detectable absorption bias is created so that the concentration of the gas to be detected can be determined. A common infrared source 16 is used so illuminate both the Signal and the Reference channels. The output of detector 5 for the Signal channel is always less than that of the detector 6 for the Reference channel irrespective whether or not there is any target gas present in the sample chamber 9. In a way similar to the Double Beam Saturation Filtering methodology, the absorption bias between the pathlengths enables the sensor to be calibrated against the concentration level of the target gas present in the sample chamber 13.

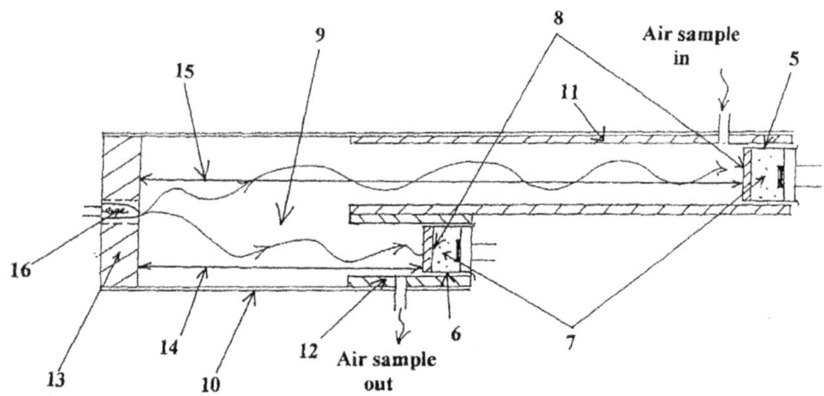

Figure 9. Optical Component Layout for the Double Beam Absorption Biased NDIR Gas Sensor.

1.3.5. Limitations of NDIR Gas Sensors

Non-Dispersive Infrared (NDIR) gas sensors have long been considered as one of the best methods for gas measurement since the 1950s. Today NDIR gas sensors are priced reasonably and widely employed. They are also highly gas specific, sensitive, fast responding, relatively stable over time, rugged, reliable and easy to maintain. Despite all these qualities, there are several inherent limitations to this methodology. First and foremost is the fact that the basic sensitivity achievable for gas detection using this method depends strongly upon the infrared absorption band of the target gas that is available. For example, for CO_2 gas which has a very strong and specific absorption band at 4.26 μ with an Absorbance value of ~0.4 (see Figure 4), detection of this gas down to a few ppm is readily achievable with a sample chamber length of ~2.0″. This is a direct result of the fact that the technical foundation for NDIR gas sensors is based upon the law of absorption known as the Beer-Lambert Law (see Equation [1] of Section 1.3.4 above). According to this absorption law, the absorption of a beam of radiation traversing a sample of gas whose concentration level is to be determined is proportional to its absorption coefficient k (equivalent to the Absorbance value for the gas), the gas concentration C and the path length L as defined by the length of the sample chamber for the sensor. While this methodology is great for detecting CO_2, it is quite a different situation when it comes to the detection of Carbon Monoxide (CO) gas. This is because the only absorption band of CO available for this methodology is at 4.65 μ and is about 10 times weaker in strength than the 4.26 μ CO_2 band. Consequently, to detect CO with the same sensitivity achievable as for detecting CO_2, the necessary path length for the sample chamber would have to be ten times longer! The same is true for the detection of many other gases which have even weaker absorption bands than CO, such as many of the indoor air polluting gases like Formaldehyde (HCHO) and many volatile organic compounds. In some situations, the sensitivity requirement for the detection of these gases has to be down to parts-per-billion (ppb) levels and the path length needed might be too long to be considered as practical using the NDIR gas detection technique.

The second limitation of NDIR gas sensors is cross-sensitivity or unwanted interference absorption. As was pointed out earlier in this Chapter, this performance limitation is also prevalent in many interactive gas sensors particularly electrochemical sensors. This cross-sensitivity is caused by the fact that the absorption band used for the detection of a target gas might overlap to some extent with one or more that belong to other gases. As an example the detection of ambient CO gas using NDIR gas absorption methodology is interfered by the presence of water vapor (H_2O) in the air since water vapor has relatively weak infrared absorption almost everywhere in the infrared spectrum. In order to correctly detect CO, the interference absorption by water vapor at 4.65 μ must be adequately compensated for. Sometimes this is easier said than done, especially for the case of CO detection where the added cost to compensate for water vapor might render this sensor too expensive for deployment in many applications.

The third limitation of NDIR gas sensors is their output drifts over time and their output dependences upon the temperature of the space in which they are located. Although it is hard to believe, sensor output instability or drift over time has been one of the major performance deficiencies for NDIR gas sensors for over six decades. But to be fair to everybody concerned, output instability over time has also been a major performance deficiency for other sensors, interactive or non-interactive, and irrespective of whatever concept or technique is used for their design. Fortunately this problem, at least for NDIR gas sensors, has finally been overcome (Wong, 2011a,b). Today NDIR gas sensors can finally be designed and built without this major performance deficiency (see Chapter 5 of this book for more details).

The fourth limitation of NDIR gas sensors is the lack of a set of simple and inexpensive procedures for re-commissioning or re-calibrating installed sensors. Again, it is fair to point out the fact that gas sensors other than NDIR gas sensors also suffer the same limitation. This limitation is especially serious and damaging because of cost issues involved for maintaining sensors accurate over time before output drift free NDIR gas sensors became recently available (Wong, 2011a,b). As a matter of fact, it is generally believed that the slow growth over the years of Demand Control Ventilation (DCV) strategy applied to buildings to save energy is caused by unpredictable and most often excessive sensor maintenance costs. Fortunately this problem has also been recently overcome. A recalibration technique for NDIR gas sensors designed using the Absorption Biased (AB) methodology (Wong, 2011b) has been advanced that can be carried out simply, remotely and rapidly without the use of any standard gases (Wong, 2011d; Wong, 2012a). The readers are referred to Chapter 5 of this book for more details.

The fifth limitation of NDIR gas sensors is their unit cost and size, but only for certain applications. Maybe it is unfair to attribute this limitation to the NDIR gas sensing methodology; instead, this limitation is maybe better attributable to the lack of adequate sensor components to support it. Let us first consider the sensor cost limitation. It has been known for quite a long time that the most reliable infrared source for this technology is the incandescent lamp or the miniature light bulb. This source is small and adequate in radiation intensity, output stability over time and unit cost. However, the glass envelope of the light bulb blocks most infrared energy beyond ~ 4 μ in wavelength, which means that many gases in the infrared spectrum (see Figure 1) cannot be measured. To detect gases having infrared absorption bands beyond ~ 4 μ, one has to resort to other kinds of infrared sources that emit longer wavelength radiation. One such source is the Micro-Electro-Mechanical-System (MEMS) device which behaves like a genuine graybody source and emits radiation much longer in wavelengths than the light bulb. But, at least for now, the MEMS source is about ten times more expensive than the light bulb, which means that an NDIR gas sensor for certain gases having absorption bands longer than ~ 4 μ using a MEMS source will have a much higher cost.

Along much of the argument given above, the spectral response of detectors used in NDIR gas sensors must match the spectral output of the source. For example, if one wishes to design an NDIR gas sensor for detecting Carbon Monoxide (CO) having its infrared absorption band located at 4.65 μ, one cannot use the light bulb as the source because of its short wavelength cutoff, and, instead, must use a MEMS source. For the detectors one must choose one with a spectral response cutoff beyond 4.65 μ. For example, a PbS photoconductive detector will not work because its spectral cutoff is at ~ 4.0 μ, which is too short in wavelength response to the relevant source radiation at ~ 4.65 μ. Instead, a pyroelectric or thermopile detector whose spectral response ranges from visible all the way to far infrared (>15 μ), a so-called "White" detector, has to be used. Therefore one of the limitations for the NDIR gas sensors has to be the availability of infrared detectors to match the absorption bands of gases using this technology. Although one can always find a suitable detector for use, the real limitation can be the much more expensive unit cost for the sensor in the end.

As referred to briefly earlier, the physical size of an NDIR gas sensor can also be limited by the weakness of the target gas absorption band available for use for its detection and the required sensitivity. Formaldehyde (HCHO) is a good example for illustration here. HCHO is an indoor air pollution gas whose presence has recently been recognized as a potential danger to people. The only available infrared absorption band for detecting HCHO lies at ~3.57 μ and is exceedingly weak with an Absorbance value < 0.04 when compared with the CO_2 gas which has an Absorbance value of ~0.4. While the sensitivity for the detection of CO_2 is adequate at ± 50 ppm, the required HCHO detection level sensitivity for workers exposed over an 8-hour workday is 0.75 ± 0.25 ppm according to a Federal standard established by OSHA in 1987. Everything being equal, if the path length required for detecting ± 50 ppm of CO_2 is around 2.0", then the required path length for the detection of 0.75 ppm of HCHO is more than 100 ft ! Needless to say, if one really is interested in the detection of 0.75 ppm of HCHO in workplaces, one can resort to a better infrared detector like a TE-cooled PbS photoconductive detector with a D^*_λ ~100 times better than the thermopile, a focused MEMS source with ~10 times higher radiation output at ~3.57 μ and a multi-bend waveguide sample chamber (Wong, 2007d) to fold path length down to ~12.0". However, once again, the limitation is really the unit cost for this sensor which would be extremely expensive if implemented with the above suggested components.

Chapter 2

A Historical Perspective

2.1. The Earliest Infrared Gas Sensing

It is generally believed that the astronomer Sir William Herschel was the first scientist to demonstrate the existence of a considerable amount of energy in the solar spectrum beyond the red end of the visible spectrum, what we now term infrared radiation, in 1800. Unfortunately, this is just interesting hearsay without any traceable documentation or proof. It is also believed that it was his son, John Herschel, that discovered the existence of absorption bands in the infrared solar spectrum in 1840. He cleverly used the sun as a source, the earth's atmosphere as the gas path, a thermometer as the detector, and a prism as a wavelength selector, to demonstrate unequivocally the presence of gases in the atmosphere-- in particular, water vapor-- and that these gases have absorption bands that can remove infrared energy from the sun. However, it was John Tyndall, noted for his Tyndall scattering effect to explain the blue sky when there are no clouds in the sky, who constructed the first spectrophotometer to measure the absorption of gases and vapors, including CO_2, water vapor and hydrocarbons (Tyndall, 1859). He observed and reported large differences in the abilities of "perfectly colorless and invisible gases and vapors" to absorb and transmit radiant heat. As a matter of fact, it is now believed that John Tyndall was the first scientist to understand the Physics behind the absorption of radiant heat from the sun by gases present in the earth's atmosphere.

2.2. The Earliest Infrared Gas Sensors

The most important NDIR gas sensors today have to be the ones that detect CO_2 and water vapor (H_2O). The application of Demand Controlled Ventilation strategy using NDIR CO_2 gas sensors to monitor the presence of occupants in buildings in order to save energy is pivotal today from an environmental standpoint. Being able to develop and produce an NDIR dew point (H_2O) sensor which is low cost, stable over time and has a long operating life (> 15 years minimum) for controlling outside cooler air via economizers, aimed again at saving energy in building ventilation, is equally important. Although NDIR CO_2 gas sensors are now readily available, this is not the case for NDIR dew point sensors. As a matter of fact, no NDIR dew point sensors have ever been successfully developed nor demonstrated until very recently (Wong, 2011e). This might be hard to believe, but it is indeed the case.

As it turns out, the very first NDIR gas sensor ever developed and demonstrated is the CO_2 analyzer. John Tyndall reported in his famous Rede Lecture "On Radiation" at Cambridge University (Tyndall, 1865) the earliest infrared quantitative measurement of CO_2 in human breath. Tyndall's experimental apparatus consisted of a long tube that he filled with various gases. This tube was equivalent to today's sample chamber for an NDIR gas sensor. To contain or trap the gas, Tyndall capped the tube with rock salt (NaCl) crystal, which is transparent to heat radiation. To create the incident light, a Leslie cube (Leslie, 1804) was heated so that it emitted radiation that traversed the tube and interacted with the gas before entering one cone of a detector called a differential thermopile. Radiation from a second Leslie cube passed through a screen and entered the other cone. A galvanometer was used to measure the voltage difference between the cones. The intensity of the two sources of radiation entering the two cones was compared by measuring the deflection of the galvanometer. The difference in the intensity of light detected indicated the absorptive power (concentration) of the gas in the tube. Amazingly, using this earliest conceived and constructed NDIR gas sensor, Tyndall and his colleagues were able to determine quite accurately the end-tidal or peak CO_2 concentration from the breaths of healthy human subjects to be between 4.66 % and 5.22 %, which agrees very well with modern findings. Figure 10 shows an artist's rendition of John Tyndall's apparatus in making infrared gas absorption experiments circa 1840.

2.3. The Period of Physics Foundation Building

Nothing much happened in the research and development of infrared gas sensing after the work of John Tyndall and his colleagues in the middle of the 19th century. The rest of the 19th century saw major advances in Physics so that by the start of the 20th century the principles behind the infrared absorption spectra of gases were known. The molecular and atomic nature of matter inferred from Avogadro's Law (Avogadro, 1810) and the work of the chemist John Dalton in the early 19th century, followed by that of Ernest Rutherford and Neils Bohr in the early 20th century on the physical makeup of the atom, provided a usable model of the structure of matter. James Clerk Maxwell's work in the mid 19th century on electromagnetism, culminating in his equations anticipating electromagnetic waves, predicted the electromagnetic nature of infrared radiation. The quantized nature of energy proposed by Max Planck in 1900 then prepared the way for Albert Einstein to round things off with his explanation of absorption and emission of radiation in 1911.

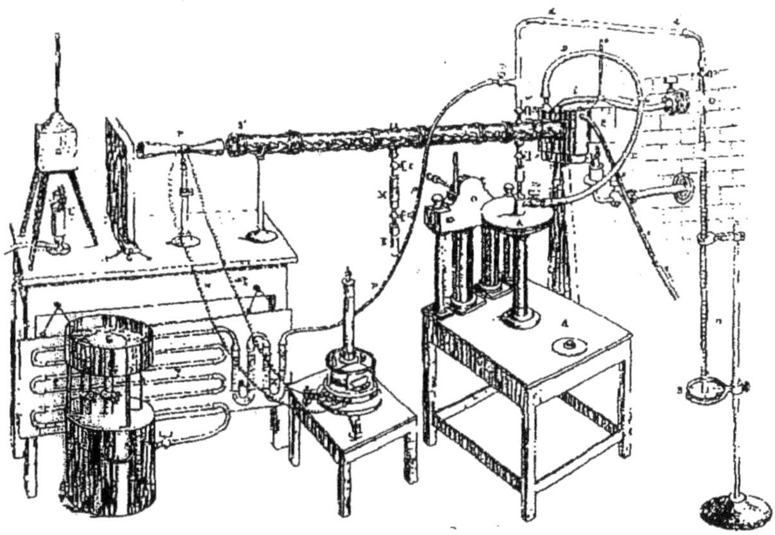

Figure 10. An Artist's Rendition of John Tyndall's Infrared Gas Absorption Apparatus circa 1840.

On the experimental front, the late 19th and early 20th century saw the observation of the absorption spectra of various gases, even the resolving of the rotational fine structure associated with certain simple molecules. Measurements were, however, very difficult to make due to the lack of suitable devices and research activities were aimed mainly at extending the boundaries of Physics rather than producing practical gas sensors. But as early as 1881 some work was done on photo-acoustic gas sensors (Bell, 1881) for the detection of Methane (CH_4). It was not until the development of non-dispersive infrared (NDIR) techniques decades later that practical infrared gas sensors became possible.

2.4. The First Practical Infrared Gas Sensors

It is quite surprising that it was the urgent need to ascertain the end-tidal CO_2 concentration level of critically ill cardiac patients that spearheaded the development of the first practical infrared gas sensors in the early part of the 20th century. In 1939, August Herman Pfund (Pfund, 1940a) developed a respiratory gas analyzer that was used at Johns Hopkins Hospital in Baltimore to measure Carbon Monoxide (CO) and Carbon dioxide (CO_2). Pfund described this thermal method of CO_2 measurement as "making possible continuous analysis without disturbing the mixture," in contrast to chemical methods that destroy the material under test. The detector cell was filled with a mixture of 3 % CO_2 and 97 % dry air. A gas sample was passed into an absorption cell and, if no CO_2 was present in the sample, the radiation was absorbed in the detector cell and the increase in temperature was measured by the thermopile. The presence of CO_2 in the absorption cell caused less radiation to be absorbed in the detector cell and therefore a smaller increase in temperature.

But the first practical NDIR gas sensor was developed by K. F. Luft in 1943 (Luft, 1943). The first version of Luft's invention is known as the Luft cell. In later years, the Luft cell was improved by replacing the capacitor detector of the device with a pair of sensitive microphones as a more balanced detector. This improved and widely adopted version was later known as the NDIR Photo-Acoustic gas sensor or the Photoacoustic Spectroscopy (PAS) gas sensor (see Section 1.3.2 of Chapter 1 above). As it turns out, PAS gas sensors became the dominant infrared gas sensors from its invention in 1943 until the end of the 1950s. Figure 11 shows schematically the components layout for the so-called Luft's device.

With reference to Figure 11, radiation coming from "a1" and "a2" is periodically interrupted through a shutter wheel "c" that is driven by motor "b". The shutter wheel "c" has slits in such a way that the radiation from both sources can pass through at the same time (or is interrupted at the same time). Behind shutter wheel "c" the radiation runs through pipes "d1" and "d2", where "d1" contains the gas to be measured and "d2" is filled with a reference gas (at half atmosphere pressure). The radiation then enters chambers "e1" and "e2", which contain – as a "reception layer" –gas that is to be measured. Both chambers are separated (in an almost air-tight fashion) from each other through a thin metal membrane "f"; this membrane forms with the insulated counter plate "g" an electrical capacitor. If the gas mixture contains the component that is to be determined, chamber "e1" receives a weaker radiation than "e2"; as a consequence, and corresponding to the interruptions by the shutter wheel "c," there is a periodic pressure difference between chambers "e1" and "e2" which can – through membrane "f" – be transformed to changes of capacitance. The membrane capacitor is followed by an amplifier "h", which allows readout at the measuring instrument "i" of the capacitance variations, and thereby the concentration of the gas that is to be determined.

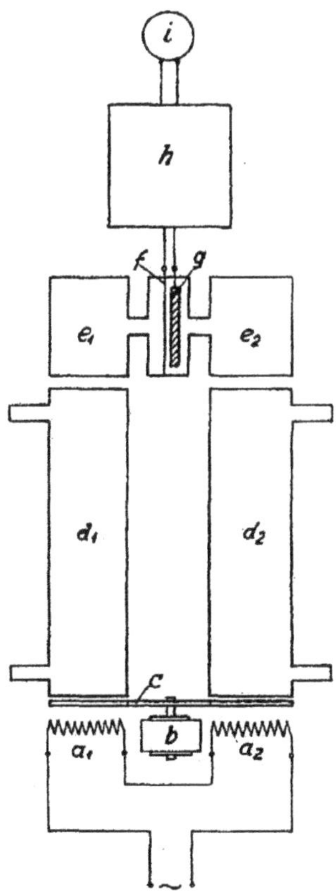

Figure 11. The Component Layout for Luft's Device – the First Practical Infrared Gas Sensor.

2.5. The Dawn of the NDIR Infrared Gas Sensors Era

As pulmonary physiology and metabolism became better understood in the early 1900s, a convenient method for measuring CO_2 in breath and avoiding the need for direct arterial measurement was keenly sought. Although the ability to measure CO_2 by infrared absorption was well understood and established at that time, early devices, dominated by Photoacoustic Spectroscopy (PAS) infrared gas sensors, were too large and impractical for clinical use. The need for better infrared gas sensors, especially those for measuring CO_2 in medical use, was clearly there, but scientist and engineers were helplessly mired and unable to break out of this quagmire due to the lack of new devices, components or simply better ideas. Consequently, nothing exciting about the development of infrared gas sensors took place during this period. Fortunately, and with literally nobody recognizing at that time, the glimmer of hope already existed even before the Second World War. As it turns out, the importance of night vision, namely the ability for someone to be able to see objects in the dark, had been deemed paramount and necessary by the military in the United States for a long time. One of the key elements required to accomplish this goal was to come up with a small and lightweight device that could select different wavelengths of light, particularly those in the infrared. The so-called dispersive elements like prisms and gratings, while able to spatially separate light of different wavelengths, were considered too bulky and heavy for use by the military to achieve their target. Instead, the military concentrated its efforts from the late 1930s on the development of a lightweight and small device that could non-spatially separate light of different wavelengths. They called this a non-dispersive narrow band-pass filter and it was designed by using spectral interference techniques similar to that for the design of the Fabry-Perot interferometer. By depositing a large number of thin films (typical 80 – 100 layers) of optical materials having different refractive indices and thicknesses one on top of the other and on one or both sides of a transparent substrate, one is able to obtain a device that can transmit selectively only a narrow band of wavelengths of light. The peak of transmission of the filter is defined as the Center Wavelength (CWL) of the filter and the narrow band of wavelengths transmitted is called the Full Width Half Maximum (FWHM) of the filter.

It is not known exactly when non-dispersive narrow band-pass filters were first successfully developed because the entire development effort for this device was classified by the military. However, it is generally believed today that this development was completed toward the end of the 1930s because weaponry taking advantage of the night vision ability was used by the Allied troops during the Second World War. It was not until the end of the 1940s, and several years after the end of the Second World War, that thin film interference filter technology was declassified by the U.S. Defense Department for public use. But the design and fabrication details of these filters were never published nor disclosed to the general public. It took the public several more years, or until 1955 or so, before these filters became available for use in the design and manufacturing of infrared gas sensors. It is therefore the middle of the 1950s that marks the dawn of the NDIR infrared gas sensors era. These sensors also got the name non-dispersive from the use of the narrow band-pass filter which, because of the fact that it does not spatially select light wavelengths like the dispersive prisms or gratings, is a non-dispersive element.

2.6. Sunrise after the Dawn

The availability of the multilayer thin film interference filters brought forth a major advance in the development of NDIR gas sensors after the mid 1950s. It was no longer the PAS gas sensors that took the center stage of the development; instead, a new class of practical NDIR gas sensors which used narrow band-pass filters for infrared wavelength selection began to catch people's attention. No longer were NDIR gas sensors designed just for the medical industry. A few started to be produced especially for use in industrial applications, like the monitoring of hydrocarbon concentration levels in petrochemical plants. But the most important advances were still made in the design of medical CO_2 sensors used in ICUs of major hospitals. This prevailed until the end of the 1960s, and it was not until then that the sun finally began to rise for today's NDIR gas sensors.

Chapter 3

The First 25 Years of Development after Dawn

3.1. Advances in Medical CO_2 Analyzers

After the declassification of the concept for narrow band-pass interference filters by the U.S. Defense Department for use by the general public in the late 1940s, and the subsequent appearance of a limited number of such filter prototypes for use by scientists at around 1955, the driving force behind the design and development for better NDIR gas sensors was still provided predominantly by the medical industry. There are at least two reasons for this. First and foremost, the medical industry at that time still was badly in need of a simpler and more convenient method for measuring CO_2 in the breaths of critically ill cardiac patients than a direct arterial measurement. The available PAS CO_2 gas sensors (see Section 1.3.2 of Chapter 1 above) were simply too bulky and awkward to use next to a patient's bed in the ICU, not to mention that their output resolution, accuracy and stability over time were far from being satisfactory. Second, although prototypes of narrow band-pass filters started to appear around that time, they were not only hard to come by, but their performance characteristics were highly unpredictable, inconsistent and outright unreliable. Therefore, from the outside looking in, it would appear that the coming of this somewhat revolutionary non-dispersive wavelength selector of light would quickly bring great advances to the world of infrared gas sensors. The actual fact is that it was not the case. Finally, other industries like the petrochemical plants, gas utilities, pharmaceutical corporations, power companies etc., had their own copious resources to either develop their own or purchase more sophisticated analytical instruments for gas measurement from the outside in order to fill their needs. The research and development for better NDIR gas sensors was generally left behind and pretty much ignored by the then contemporary scientists.

The first NDIR gas sensor surfaced in the early 1960s that was worthy of note was Beckman Instruments' Model LB-I medical CO_2 analyzer. This analyzer was among the first to utilize a narrow band-pass filter centering around the 4.26 μ CO_2 absorption band for its design. It was therefore one of the first genuine NDIR gas sensors, not a PAS gas sensor, offered for sale to the general public. Its operation is based upon a Single Beam Configuration Design (see Section 1.3.4 of Chapter 1 above). Due to this configuration, there are many factors, other than the target gas, that can lead to a change in the detected signal, causing a measurement error. These factors include changes in the infrared source output, the filter spectral transmission and detector responsivity and spectral response with both time and temperature, not to mention errors that might be caused by signal processing electronics and fouling of sensor optics. Still, because of the fact that the Beckman LB-I was much smaller than a PAS CO_2 sensor and easier to use beside a patient's bed, it was welcomed by the medical industry and widely used in ICUs of many hospitals when it was first made available. But it did not take long before people started to complain about its overall poor performance, particularly about its output instability over time which caused serious errors in patient CO_2 measurements. The problem grew to be so bad that it became a mandatory procedure for hospitals to re-calibrate the LB-I analyzer for output accuracy every 30 days !

It was Hewlett-Packard's Model 47210A medical CO_2 analyzer, otherwise known as the "Capnometer," that made the greatest impact, not only on infrared gas sensor technology per se, but also on the legendary sensor performance characteristics which were considered the "gold standard" for the medical CO_2 analyzer industry until around the end of the 1980s. The story of the development of the Capnometer (Jaffe, 2008) is part of the overall story of Hewlett-Packard Company's entrance into medical devices in the late 1960s. It is also a story of adapting technology intended for the automotive industry to medical applications. Since one of the authors (JYW) was the Capnometer's project manager when the idea was first conceived in the Hewlett-Packard Laboratories (HPL) in Palo Alto, California in late 1968, and the story itself is so compelling and interesting to tell in so many aspects, with the readers' assumed indulgence, the Capnometer story will be recounted here in detail below.

3.2. The Capnometer Story

In early 1967, Hewlett-Packard Laboratories (HPL) won a major contract worth over $ 1 M from the General Motor Corporation (GMC) to develop a small, compact and rugged multi-component exhaust gas analyzer for the company's automotive assembly lines. The gas components required to be measured were Hydrocarbons represented by Methane (CH_4), Carbon Monoxide (CO), Carbon Dioxide (CO_2) and Nitric Oxide (NO). The measurement accuracy required was, respectively, CH_4 (± 50 ppm), CO (± 50 ppm), CO_2 (± 100 ppm) and NO (± 50 ppm), and they had to be accurately determined within a time period of 10 seconds (0 – 90 %) or less. Prototypes for this multi-component auto exhaust gas analyzer had to be ready for tests within a period of one year. Probably because there were only a limited number of General Motor automotive assembly plants in the United States at that time, the required unit production cost for the multi-component exhaust gas analyzer was not even specified explicitly in the signed development contract.

Looking back today with 20/20 hindsight, the exhaust gas analyzer proposed by HPL for the automotive assembly line must be considered as an absolute impossibility purely from the standpoint that the technology for the development of such a sensor was not available or even ready at that time. Even with the most optimistic fact that the various narrow band-pass filters required for the detection of the specified gases were available and their performance could be proven to be reliable, the show stopper would still have to be the ruggedness of the sensor because a mechanical chopper still had to be used to provide the needed AC signal processing data. Furthermore, the size of the final sensor would be another major obstacle to overcome because whether or not such a fragile and bulky multi-component exhaust gas sensor could easily be fitted for use at the end of an automobile's tail pipe was clearly questionable. Notwithstanding, the supervisor of JYW at HPL at that time was an exceptional individual who survived the Auschwitz concentration camp during the Second World War and who could never say "NO" to anything. Furthermore, he was then part of a R&D team at HPL that had successfully grown the epitaxial GaAsP and GaP semiconductor materials for fabricating, respectively, the red and green light emitting diodes (LEDs). As a matter of fact, it was he that was instrumental to winning the GMC analyzer contract for HPL.

But in the end one can never get blood out of turnip. Not long after getting the GMC contract, the HPL project team soon decided that a conventional design approach for the GMC exhaust gas analyzer would not work and a new and revolutionary scheme had to be developed to take its place. This new idea was to combine the infrared source and possibly the filter required for gas detection together into an infrared light emitting diode (IRED). Since one can readily pulse an IRED at any frequency, the mechanical chopper would no longer be needed to work the sensor and all the problems associated with it would disappear. Riding on their recent technical accomplishment of successfully growing epitaxial GaAsP and GaP semiconductor materials for making red and green LEDs, the GMC sensor project team turned their attention to the growth of another III-V compound, namely, an epitaxial ternary Indium Arsenide Antimonide ($InAs_{1-x}Sb_x$) semiconductor material for making infrared light emitting diodes (IREDs). Since the spectral cutoff for the semiconductor material of InAs and InSb is ~3.3 μ and ~5.6 μ, respectively, if the growth of the epitaxial ternary compound $InAs_{1-x}Sb_x$ could be achieved, then it would be possible to fabricate individual IREDs to emit at radiation wavelengths covering the respective absorption bands for the four required gases. For CH_4 detection, an IRED could be fabricated that emits at ~3.4 μ using the pure binary InAs material. For CO_2 detection, an IRED could be fabricated that emits at ~4.3 μ using the epitaxial $InAs_{0.92}Sb_{0.08}$ ternary material. For CO detection, an IRED could be fabricated that emits at ~4.60 μ using the epitaxial $InAs_{0.88}Sb_{0.12}$ ternary material. Finally, for NO detection, an IRED could be fabricated that emits at ~5.20 μ using the epitaxial $InAs_{0.88}Sb_{0.12}$ ternary material. For the infrared detector that could cover the spectral response range from ~3.4 – 5.20 μ, either an InSb PC or PV detector could adequately do the job. For some optimistic members in the GMC analyzer project team, a better infrared detector might well be just the thermopile detector.

This new and revolutionary approach for the design of the 4-component exhaust gas analyzer proposed at around 1967-1968 at HPL was brilliant and unprecedented. If this design scheme could be proven to be successful, not only could a very small multi-component gas sensor be realized, but also this sensor would be very rugged and work without the need for any moving parts such as the mechanical chopper. Although the unit cost for this type of gas sensors could be a problem in the beginning, their actual realization would no doubt have made an enormous impact on the infrared gas sensor industry at large, and drastically change the scenery of what we are seeing today.

It took the GMC sensor project team almost 8 months working as hard as they could to come up with the first epitaxial ternary $InAs_{1-x}Sb_x$ materials good enough for making the first batch of such IREDs. But the results were most disappointing. It was quite well known even at that time that in order to fabricate efficient LEDs or IREDs, in other words PN junctions or diode devices that emit bright light, the quality of the material making them must be excellent and should have as low a density of recombination centers in it as possible. A recombination center simply acts as an electrical carrier (electron or hole) trap and the presence of a whole bunch of them would eliminate the carriers, namely electrons and holes, before they have a chance to recombine to generate a photon. Thus, if the density of recombination centers is high, the efficiency of light generation would be low. One way to quantitatively characterize the quality of a semiconductor material for making LEDs or IREDs is by measuring the external quantum efficiency value (η_e) of the PN junction device fabricated from it. When $\eta_e = 1.0$, it means that for every carrier (electron or hole) injected into the PN junction, one photon will be generated. When $\eta_e = 0.1$, it means that one photon will be generated for every 10 carriers injected into the PN junction. For ordinary red LEDs fabricated out of epitaxial ternary GaAsP material, $\eta_e = \sim 10^{-1}$ to 10^{-2}. Thus, by measuring η_e, one could readily determine the quality of the semiconductor material employed to fabricate LEDs or IREDs.

The epitaxial ternary $InAs_{1-x}Sb_x$ materials grown at HPL in the beginning were indeed of very poor quality. The measured values for η_e from the IREDs fabricated out of them were around 10^{-6}. In the beginning, there was hardly any infrared radiation that could even be detected from them at all despite using the most sensitive detector available (double-stage TE-cooled InSb PC detector with a D^* of $\sim 10^{11}$ $cmHz^{0.5}W^{-1}$). Subsequent lots of epitaxial ternary $InAs_{1-x}Sb_x$ did not yield much improvement in the measured η_e, only by a factor of ~10. Intense research followed trying to find out the reason why the density of recombination centers was so high for this epitaxial ternary material that led to such low measured values of η_e. It was soon found out that the culprit was the physical lattice mismatch between the different compositions of the epitaxial ternary compounds grown on the InAs binary substrate and most of the recombination centers were formed right along the binary/ternary boundary of the epitaxial ternary $InAs_{1-x}Sb_x$ material. Needless to say, without better materials to make the IREDs, the multi-component exhaust gas analyzer project was doomed to failure.

But the supervisor of JYW at that time never learned to say "NO" and always behaved as stubborn as a mule. As a last resort for generating enough infrared radiation from the IREDS to make the sensor, he decided to cool the devices down to liquid nitrogen temperature. The results were truly astonishing. By cooling the IREDS down to ~77 ^{o}K from ambient temperature, there was a three orders of magnitude improvement in the η_e and all four components of the gases could now be adequately measured. But the one-year project time was running out and there was hardly any doubt that the exhaust gas analyzer could not operate properly at ambient temperature, but rather had to be cooled with liquid nitrogen, and that operating the gas sensor with cooling with liquid nitrogen, while it would meet the performance specifications, would be unacceptable to the project sponsor. The ending of this GMC exhaust gas analyzer project would have been a total disaster for the project team if not saved by one of the lawyers of Hewlett-Packard Company, a member of the project team itself !

When this lawyer looked carefully through the signed contract for this project, he observed that the contract never said that the analyzer had to operate at ambient temperature. Apparently it was an oversight by the project sponsoring team at GMC when the contract was signed. Therefore, the contract terms were met by HPL when the analyzer was delivered to GMC and its demonstration worked well except for the specification of its operating temperature. General Motor Corporation was not very happy, nor was Mr. Hewlett, when he was subsequently told about what happened. Nevertheless, the fact remained that HPL spent over \$ 1M of GMC's money in learning how to advance the technology for infrared gas sensors.

Again, looking back today with a 20/20 hindsight, the \$ 1M spent by General Motor Corporation in 1967 trying to develop a multi-component exhaust gas analyzer for its automobile assembly lines was money well spent at that time even though it did not result in success from the standpoint of advancing the overall technology for infrared gas sensors. This is because from the remnants of an automotive program, the idea of a medical CO_2 sensor was born. As it turned out, Hewlett-Packard Company had already entered the medical field with the purchase in 1961 of Sanborn Company, a cardiac paper manufacturer in Waltham, Massachusetts. Bill Hewlett's vision for Sanborn was to be the foundation of the future HP medical division. The first project he commandeered at HPL to realize his vision shortly after the acquisition of Sanborn was the development of the well-known 8-wavelength ear oximeter (Hayes et al., 1976). By the time the GMC exhaust gas analyzer project beam wrapped up their work at HPL, Mr. Hewlett saw another opportunity to further his vision. He was aware of the fact that by then the HPL GMC analyzer team had the theoretical understanding of infrared absorption of all the gases. He therefore quickly mandated the development of a medical CO_2 sensor and appointed JYW to spearhead the development.

While it was indeed an honor and a great career opportunity for JYW to be appointed by the top man of the company to head up this project, it did not take long for him to realize that this could also be a career-wrecking trap. The reason was that most of JYW's colleagues at HPL assumed at that time that most or at least some of the infrared absorption technology developed in the GMC exhaust gas analyzer project could be used to advantage in the medical CO_2 sensor development. But actually this could not be any farther from the truth. As it turned out, the $InAs_{0.92}Sb_{0.08}$ IRED could not be used for the medical CO_2 sensor as its η_e was still way too low if not operated at liquid nitrogen temperature (77 ^oK). Meanwhile the medical CO_2 sensor must be operable at ambient temperature by the patient's bedside. Without the use of the CO_2 IRED, all the problems earlier associated with the bulky mechanical chopper with regards to sensor ruggedness and size reappeared. But even worse than that, while the measurement response time required for the GMC exhaust gas analyzer was ~10 sec (0 – 90 %), the measurement response time for the medical CO_2 sensor had to be 100 ms or less. With using the CO_2 IRED as the source, there should be no problem as the IRED is readily pulsable at any frequency. But with only the mechanical chopper available, this fast response requirement posed yet another major problem that one had to contend with. Furthermore, the size of the medical CO_2 sensor had to be even smaller than the GMC exhaust gas analyzer because it had to be accommodated befittingly onto a patient's respiratory airway with the use of an airway adapter. The only saving graces were that one needs only to deal with the measurement of one gas, namely CO_2 with an extremely strong and specific absorption band at 4.26 μ, and that the sensor cost did not seem to be a problem that one had to be concerned with.

At the time when Hewlett-Packard's medical CO_2 sensor project started at the HPL, the most widely used medical CO_2 sensor in hospitals was the Beckman LB-II, a somewhat improved and smaller version of the earlier Beckman LB-I instrument. The LB-II used a Double Beam

Configuration Design (see Section 1.3.4 in Chapter 1 above) instead of a Single Beam Configuration Design (see Section 1.3.4 in Chapter 1 above) used by the LB-I. Unfortunately, sensor output instability over time was still the main complaint by LB-II users and mandatory recalibration of the sensor every 3 months was still rigidly enforced in most hospitals using it. This performance drawback for the LB-II was so well-known among hospitals that it even impacted the final design for HP's medical CO_2 sensor. Besides size, ruggedness, fast response time and measurement accuracy, sensor output stability over time had become the single most important performance goal for the sensor.

For the project leader of HP's medical CO_2 sensor project, the first three months were utter frustration and despair. Designing the sensor around an efficient CO_2 IRED operable at ambient temperature would probably make the project manageable and the availability of a reliable narrow band-pass interference filter with a CWL=4.26 μ and a FWHM = 0.14 μ would certainly make the design task easier. But unfortunately both of these crucial components were out of reach for this difficult project. The situation was very much like what the GMC exhaust gas sensor project team faced at the beginning of that project. Conventional thinking would not work and had to be thrown out and a new and revolutionary mindset was needed in order to have any chance of success at all, no matter how small. It was with this pioneering and challenging spirit in research and development that the project leader finally went to work.

Knowing full well that the most important performance goal for HP's medical CO_2 sensor was output stability over time along with the small size, fast response time and accuracy requirements but without the services of either the CO_2 IRED or a reliable narrow band-pass filter for the detection of the gas, the main challenge was to come up with a new and innovative gas measurement technique that could meet this goal. After several months of intensive soul searching in Physics, the project leader finally recognized the fact that the 4.26 μ CO_2 absorption band was exceedingly strong and, if designed properly, the CO_2 gas itself could serve as its own filter for gas measurement. He advanced what he then called the "Negative Filtering" technique for gas detection. This technique only worked for measuring gases with very strong infrared absorption bands such as CO_2. The basic technique was based upon the Double Beam Configuration Design (see Section 1.3.4 of Chapter 1 above) for eliminating all non-gas phenomena affecting the sensor by taking the ratio of the signals from the Signal and the Reference channels as its output. The conventional approach to create the Signal and Reference channels for the Double Beam Design was to use respectively two narrow band-pass filters with the CWL of one at the gas absorption wavelength and the other with a CWL at a non-absorbing neutral wavelength. But for HP's medical CO_2 sensor project, the narrow band-pass filters available at that time were considered not reliable enough for use in medical instruments. Fortunately the new and innovative "Negative Filtering" gas measurement technique solved this problem.

Without the use of narrow band-pass filters, the Reference channel was created by interposing in front of the detector a small thin cell about 6 mm thick containing 100 % of CO_2 gas roughly at atmospheric pressure. The windows of this cell were chosen to be sapphire which is transparent from the visible all the way up to ~6.0 μ in the infrared. Because of the fact that CO_2 has such a strong absorption band at 4.26 μ, the CO_2 cell drastically diminishes any influence of CO_2 gas present in the sample chamber of the sensor. In other words, the detector signal from the Reference channel hardly changes whether or not there is CO_2 gas present to be measured. Its behavior is similar to the Reference channel of a conventional Double Beam Design using a neutral narrow band-pass filter with a CWL off the absorption band of the CO_2 gas.

To create the Signal channel for the medical CO_2 sensor utilizing the "Negative Filtering" methodology, a similar small thin cell was interposed in front of the detector except that the cell was now filled only with 100 % dry nitrogen at approximately atmospheric pressure. Thus, if there was any CO_2 gas present in the sample chamber, the Signal detector output would decrease due to the CO_2 absorption. In fact, the Signal detector output would be modulated up or down according to the concentration of CO_2 gas present in the sample chamber. Its behavior is similar to the Signal channel of a conventional Double Beam Design using a narrow band-pass filter with a CWL coincident with the wavelength of the absorption band of the CO_2 gas.

In order to simulate the conventional Double Beam Design for HP's medical CO_2 sensor, the two cells, namely, one containing 100 % CO_2 and the other 100 % dry Nitrogen, were alternately placed in front of one and the same infrared detector via the use of a small and specially designed aluminum magnetic rotor at a rate of 40 Hz. As discussed earlier, the Signal channel was formed when the 100 % N_2 cell was aligned between the source and the detector and the Reference channel was formed when it was the 100 % CO_2 cell that assumed that position. The extremely well designed miniature magnetic rotor that tightly houses the two cells referred to above took the place of the bulky mechanical chopper in the conventional Double Beam designed infrared gas sensor. Meanwhile the ratio of the Signal channel output over the Reference channel output was processed to generate the concentration of the CO_2 gas present in the sample chamber.

To complete the design for HP's medical CO_2 sensor, a heated thick film blackbody radiator was used as the infrared source. This was the same printer head source used in Hewlett-Packard's well-known and reliable thermal printer available for sale to the general public in the early 1960s. This particular printer head can operate in air without damage and has a typical life span of more than 5 years under continuous use. Since the magnetic rotor containing the gas cells was driven at 40 Hz to generate the AC signal required for the sensor signal processor, the heated thick film source for the sensor was simply run at DC. In order to meet the high accuracy and fast response time requirements for the sensor, a room temperature PbSe photoconductor was used as the detector. As a general guideline for the design of good performing infrared gas sensors, it is always wise to limit the spectral band of radiation seen by the detector and that is the reason why the use of a narrow band-pass filter to detect the gas, if available, is so effective. In the present case there was no narrow band-pass filter at ~4.26 μ with a FWHM = ~0.14 μ to provide an isolated spectral band for the detection of CO_2. Fortuitously and maybe by luck, a less pronounced spectral band had actually been isolated for this purpose between the rising portion of the excitance curve for the thick film blackbody source (~250 °C) and the spectral response cutoff of the PbSe photoconductive detector at ~4.5 μ.

It took a little more than two years to complete the project. When the prototypes were finally ready for performance testing, the results were so good that they were almost unbelievable. All the performance expectations--including accuracy, fast response time and small size (including the airway adapter for interfacing with the patient's airway)--were met with room to spare. But the most astounding performance accomplishment was its output stability over time, which of course had been the single most important performance goal to achieve. This surprising result was indeed somewhat puzzling to every member of the project team and even so to most of the staff working at HPL at the time. While nobody including the project leader clearly understood at that time the reason why the output of the sensor was so stable over time, looking back today, the reasons can be determined and are given below.

First of all, a lot of credit must be given to the project team members responsible for the design of the magnetic rotor that rotates the wheel carrying the gas cells to put them alternately in front of the detector at a frequency of 40 Hz. The rotor was so well designed that not only was it extremely rugged, the wheel it rotates was so even and repeatable in position that hardly any output signal jitter was ever observed when the sensor was in use. That was one of the reasons why the performance accuracy for the sensor was so remarkable (± 25 ppm @ 100 ms response time). A lot of credit must also go to the project team members responsible for the fabrication of the two gas cells, namely the 100 % CO_2 and the 100 % N_2 cells. Although they were very expensive--costing ~US $ 200 each to make-- their performance over time was definitely worth their cost. These cells had custom designed kovar bodies with sapphire windows metallically braised on them from both sides. A special high temperature solder was used to seal the gas inside the cell via a custom designed sealing procedure developed especially for this operation. The end result was that the gas cells were extremely gas tight and stayed leak free for years on end. They were one of the key reasons why the output of the sensor stayed drift free for long periods of time.

But the main reason why the output of HP's medical CO_2 sensor stayed drift free for very long periods of time without the need for recalibration was the use of the "Negative Filtering" technique for its design. Except for the gas cell containing either 100 % CO_2 or 100 % N_2 respectively for creating the Signal and Reference channels, the two channels were almost completely identical. By taking the ratio of the Signal channel output over the Reference channel output, all the non-gas phenomena--including changes in the source output, changes in the detector spectral response or responsivity, or sample chamber (airway adapter) optics--were genuinely eliminated. After all, both Signal and Reference channels share the same source, the same detector and the same airway adapter (sample chamber). Only the presence of CO_2 gas in the optical path will affect the ratio. The conventional Double Beam Configuration Design (see Section 1.3.4 of Chapter 1 above) for an NDIR gas sensor is vulnerable to spectral output changes for the source as the Signal and Reference channels will be affected differently because they have different narrow band-pass spectral filters. Looking back in time, it was truly amazing that the Hewlett-Packard medical CO_2 analyzer, otherwise known as the Capnometer, was able to achieve the feat of output stability over time for years on end over five decades ago !

But the telling of the Capnometer story would not be complete without recounting how the name for this sensor, namely "Capnometer," came about. At the time the project team was developing the sensor at HPL, the sensor was assigned a model number right from the beginning because it was Mr. Hewlett's "pet" project. It was named Model 47210A. By the time the development of the sensor was completed at HPL, everybody referred it as the Model 47210A medical CO_2 analyzer. When the project was transferred over to HP's Waltham Medical Division in 1970 for its final development for production, the marketing people did not like the long Model number for addressing the instrument. The reason was as follows. After the marketing people witnessed the performance of the prototype, they knew right away that this analyzer would take over the medical CO_2 sensor market when it became available for sale. At that time everybody knew about the Beckman LB-I and LB-II CO_2 instruments even though they had a bad reputation because of their poor performance. But LB-I and LB-II were so simple and easy to refer to, versus the Model 47210A for HP's medical CO_2 analyzer that something had to be done to simplify or shorten its name. But the problem was that HP at that time was a very conservative company. The company had never before called their instruments or products by nicknames other than their Model numbers. So to come up with a shortened name for HP's medical CO_2 analyzer was indeed a very tall order to fill.

This situation was resolved when the project leader came up with the name "Capnometer." The reason for this name was as follows. The word "Capno" is a Greek word meaning smoke. But because of the fact that smoke normally comes from a fire when there is a fire burning, there would be plenty of CO_2 gas produced. Since the word "meter" means a very accurate and precise instrument, combining the Greek word "Capno" with the word "meter" results in the combination "Capnometer" which suggests that the sensor is a very accurate instrument for measuring CO_2 gas. With the name ready to go, it still took the project leader over three months to convince his supervisor to arrange a meeting with Mr. Hewlett at corporate for approval of the new name. Fortunately, Mr. Hewlett apparently was in a very good mood during the short meeting and the name "Capnometer" was approved for calling HP's Model No. 47210A medical CO_2 analyzer !

The sale of the Capnometer to the general public started in 1975. The sales price for the first batch of units was US$ 6,500.00 each. The Capnometer remained as one of the most successful instrument of its kind and despite its very expensive price tag, it was estimated that over 50,000 of them were manufactured and sold. It served as the "gold standard" for medical CO_2 sensors for almost two decades after its introduction in 1975 !

3.3. The Development Period 1955 - 1980

As discussed earlier in this chapter, the driving force behind the development for better NDIR gas sensors during this period was provided predominantly by the medical industry. The introduction of Beckman Instruments' LB-I and LB-II medical CO_2 analyzers in the 1960s, the Hewlett-Packard Capnometer in 1975, and the Siemens-Elema AB CO_2 sensor in 1976, were prominent examples. It must be noted that practical and reliable narrow band-pass interference filters were not really available for the development of NDIR gas sensors until the early 1970s. Also there was never any doubt that development work was on-going during this time period except that scientists and engineers working for private companies did not publicly disclose any of their significant findings or results. Still, some of the development work on NDIR gas sensors could be found in issued patents, particularly from the United States Patents and Trademark Office (USPTO). The following are some of the noteworthy development work as disclosed in issued USPTO patents during this time period.

In 1974, Sternberg et al. (Sternberg et al., 1974) disclosed a method and system for the infrared analysis of gases. In one of the embodiments of this disclosure, a single path instrument includes a source of infrared energy, a detector for the energy, a sample cell for the gas mixture positioned between the source and the detector, and a filter wheel having a plurality of filters and a source blocking device positioned between the source and the detector for sequentially interposing the filters and the source blocking device between the source and the detector. Means are provided for amplifying the output signal of the detector and for processing the signal to provide a direct readout display which indicates the concentration of the gaseous components being analyzed. This United States Patent and Trademark Office

(USPTO) patent disclosed for the first time the methodology of a multi-component infrared gas sensor utilizing different narrow band-pass filters mounted on a chopper wheel for the detection of different gases according to the wavelengths of their infrared absorption bands.

Also in 1974, Burch et al. (Burch et al., 1974) reported the first successful operation of a dual cell non-dispersive gas analyzer. They described their gas analyzer as having a beam of radiant energy, chopped at a suitable frequency, and limited by filter means to a suitable portion of the absorption spectrum of the target gas, which is passed through a sample region containing an unknown quantity of target gas to be analyzed. Selector cells are alternately inserted into the beam at a rate substantially lower than the chopping frequency. One selector cell contains a quantity of the gas species to be analyzed at high pressure and the other contains a quantity of the same gas at low pressure. A photo detector converts the radiant energy variations to electrical signals which can be analyzed to indicate the quantity of target gas in the sample region. A 20/20 hindsight reveals the fact that the invention of Burch et al. is quite similar in concept to that of the Hewlett-Packard Capnometer. Whereas the "Negative Filtering" technique of the Capnometer utilizes a cell containing 100 % of the target gas (CO_2) and a cell containing 100 % of an inert gas (N_2) respectively in the Signal and Reference channels, the dual cell analyzer of Burch et al. uses a cell containing the target gas at high pressure and a cell containing the same gas but at low pressure respectively as the Signal and Reference channels. The signal processing methodology for these two gas measurement techniques were the same in that each rely upon using the ratio of the Signal channel output over the Reference channel output as the processing parameter.

Also in 1974, Blau (Blau, 1974) disclosed in principle the first Double Beam Configuration Design (see Section 1.3.4 of Chapter 1) for NDIR gas sensors. In one of his disclosed embodiments for the invention, gas in a sample region is analyzed in a non-dispersive system for the presence of a particular gas of interest by cross-correlating the absorption spectra of the sample and a specimen of the gas of interest. Radiation is directed through the sample region and a filter for isolating a narrow band of frequencies within the absorption band of the gas of interest, to a radiation detector. The radiation is passed through cells that provide the total energy transmitted at the line or lines of interest and a reference energy transmitted by a region adjacent to the lines of interest but unaffected by changes in absorption in the sample region. To translate Blau's invention into layman terms, basically he created two channels, one Signal and the other Reference. By using the correct filter, the Signal channel output is affected by the target gas in the sample region. The selection of another filter slightly away spectrally from the absorption lines of the target gas makes the Reference channel output unaffected by the presence of the gas in the sample region. Thus, just like the "Double Beam Configuration Design", by taking the ratio of the outputs of the Signal and Reference channels as the processed signal for the sensor, all the non-gas phenomena, such as changes in the source and detector characteristics over time, can be substantially eliminated.

Again in 1974, French (French, 1974) disclosed a dual wavelength type moisture gauge which is basically an infrared gas sensor for the detection of water vapor (H_2O) utilizing a Double Beam Configuration Design. In French's invention, an infrared source provides a beam of radiation having a constant intensity and an output spectrum that spans at least one absorption band of water vapor (H_2O). A detector receives radiation transmitted or backscattered from the material whose moisture is to be determined. Within the detector assembly, transducer or converter means are provided to convert the received radiation to two electrical signals, one signal being representative of the intensity of radiation received in a spectral band highly sensitive to absorption by water vapor and the other signal being representative of the intensity of radiation received in a spectral band near that of the first but relatively insensitive to the absorption by water vapor. But taking the ratio of the two electrical signals, the amount of water vapor in the material can readily be determined.

In 1975, Pembrook et al. (Pembrook, 1975) disclosed an electronically compensated rotating gas cell analyzer. In this disclosure, a beam of radiant energy is passed through a rotating gas cell which alternately inserts optically transmissive chambers into the path of the beam. One chamber contains a quantity of gas of the species to be analyzed at a pressure that produces sharp absorption lines. The other chamber is provided with either a quantity of gas at a pressure that produces substantial line broadening or a neutral density attenuator. In either case the chambers are made to have substantially equal average transmissivity over the spectral band of interest. The beam is then passed through a sample region in which the gas species concentration is to be measured. The beam is then detected by a photodetector. A filter, usually associated with the photodetector, limits the radiant energy response to a spectral region where the gas species to be measured displays suitable absorption lines. An electronic system amplifies the photodetector output and demodulates that signal component which occurs at a frequency corresponding to the cell rotation rate. In the absence of the particular gas in the sample region the two cell chambers have equal average transmissivity, and there should be no such component. However, if a quantity of the gas species to be measured is present in the sample region, a rotation rate signal component will be present and will be in proportion to the total quantity of that gas species in the sample region.

In summary, the development activities and accomplishments for infrared gas sensors during the period 1955 – 1980 are quite difficult to ascertain due to the lack of any prominent books or research articles appearing in published scientific journals. It is speculated that this is because scientists and engineers working in this field at that time were not allowed to publish any significant findings or results due to competitive reasons and concern for trade secrets. Because of the lack of published work, the only source for publicly available information is relevant USPTO patents. This somewhat unorthodox behavior for research scientists and engineers in this field still persists even until today.

Chapter 4

The Next 30 Years of Development

4.1. The Growing Need for NDIR Gas Sensors

The period from the early 1980s to 1990s witnessed a rapid growth of diverse applications in the use of NDIR gas sensors. Besides the medical and research industries which were the mainstay for NDIR gas sensors throughout the 1970s, other industries, such as mining, oil and gas production, HVAC, petrochemical, pharmaceutical, cosmetics, and diverse manufacturing of all manners of goods etc. where safety, efficiency, better living, personal health and comfort have become invariably of the utmost importance, started to demand more and more NDIR gas sensors, beginning in the 1980s. Their demand did not focus just on the general availability in numbers and their detection of diverse gases, but also on the size, ruggedness and cost for these sensors. Throughout the 1970s NDIR gas sensors, especially those that implemented the Double Beam methodology of gas measurement, were bulky, costly, relatively heavy and included moving parts such as mechanical light choppers.

Beginning in the 1980s, coinciding with the availability of low-cost microprocessor platforms, researchers and developers of NDIR gas sensors concentrated on new sensor designs that were compact, light-weight and possessed no mechanical moving parts or focusing optics. This marked the evolution point of infrared gas sensors from mechanical-optical systems to electronic systems. In addition, significant cost reductions were achieved. But looking back into that period of time when the need for gas sensors was growing leaps and bounds, the focus of attention was not just directed to NDIR gas sensors per se. The demand for gas sensors also ushered in the development of gas sensing technologies other than NDIR. The most notable technologies were electrochemical, solid state conductivity (TGS) and photo-ionization. So at that time NDIR gas sensors at large were also under heavy pressure from gas sensors of other technologies to compete and excel, particularly against electrochemical gas sensors which, despite their relatively poor performance, were very low cost and tiny and could readily detect a large number of gases down to ppm levels.

4.2. Noteworthy Developments in the Period 1981 – 1990

The noteworthy inventions during the period of 1981 – 1990 that were disclosed in issued USPTO patents are summarized briefly below.

In 1981, Eden (Eden, 1981) disclosed the use of a thin film of vanadium oxide, switchable between a high reflectance state and a high-transmission state, for switching a light signal from one direction to another. The thin film is alternately heated and cooled by thermoelectric junctions placed in thermal contact with the edges of the substrate on which the thin film is deposited. This invention represents the first beam switching device without any mechanical moving parts. Unfortunately, it cannot be readily used for switching infrared energy beams as there is always an infrared source present at the same switching frequency due to the heating or cooling pulse needed to switch the optical transparency state of the thin film.

In 1985, Maiden (Maiden, 1985) disclosed an apparatus for non-dispersive optical determination of the concentration of an identifiable gas, such as a hydrocarbon, in a mixture such as atmospheric air. The apparatus comprises a sample chamber to contain a sample of the mixture, a source of IR radiation at a wavelength absorbed beamed through the sample chamber to a detector and transducer means to change the gas pressure within the chamber to determine the change in radiation intensity detected at different pressures. One wall of the sample chamber is formed by a flexible diaphragm driven by a voice-coil type transducer whereby the pressure is switched repetitively between the values above and below atmospheric. Atmospheric gas is admitted to the chamber by a porous plug which restricts flow to a low level permitting a pressure change in the chamber. Absorption by the gas relates the intensity of the source radiation I_S and measured radiation I_M. By measuring intensity at different pressures involving a change Δ_p and a fractional change in intensity of measured radiation $\Delta I/I$, the gas concentration can be determined.

In 1986, Wong (Wong, 1986) extended the "Negative Filtering" methodology he pioneered in the late 1960s for the measurement of CO_2 gas used in the legendary Hewlett-Packard medical Capnometer, by inventing a self-calibrating CO_2 analyzer for medical purposes. In addition to the two cells used in the Capnometer, one containing 100 % CO_2 in the Reference channel and another one containing 100 % N_2 in the Signal channel, a third cell containing ~5 % of CO_2 gas balanced with 100 % N_2 is additionally deployed in the standard channel. A very intricate two-wheel chopper with mirror arrangements is used to establish the standard channel independent of the target gas present in the sample chamber so that the analyzer can always readjust the measured ratios in order to recalibrate itself to this standard. Although the methodology for this invention was indeed novel and sound, it was heading in the wrong direction as the development trend for NDIR gas sensors at that time was toward simpler, smaller and lower cost sensors instead of more complicated and expensive instruments, even if they possess superior performance capabilities.

Also in 1986, Yamada (Yamada, 1986) disclosed a gas analyzer with a compact cell structure. The object of the invention is to provide an extremely simple and yet effective improvement to a single beam type infrared gas analyzer. This gas analyzer has a light source and a sample cell aligned with the light source and through which the light from the light source passes and received by a detector in series. The sample cell is a simple hollow cylinder open at both ends. The light source is fitted into one end and sealed with a window while the detector is fitted to the other end also sealed with a window. The sample cell has a rotatable chopper for interrupting the light from the source to the detector. A gas inlet is also provided to the sample cell away from blocking the light passing through it.

In 1987, Aoki et al. (Aoki et al., 1987) disclosed a gas analyzer with means for varying the angle of incidence of light on interference filter. A gas analyzer provided with apertures on the optical paths from a light source through a cell to sensors and multi-layer interference band-pass filters on the respective optical paths. The structure makes it possible to change the distance between each sensor and the corresponding aperture, or to change the size of the aperture for changing the maximum angle of incidence of light on the filters. Effectively the central wavelengths of the filters are thereby changed as far as the light passing through the apertures and sensed by the sensors is concerned.

Also in 1987, Wong (Wong, 1987) disclosed a non-dispersive gas analyzer having no moving parts. In this invention, a multi-component non-dispersive gas analyzer of a type typically used to measure the concentrations of gases present in automotive exhaust emissions and in medical breath analyzers has no moving parts and employs electrically operated means for effectively inserting and removing a reference cell from the optical path and for selecting a particular filter to determine momentarily the wavelengths of radiation being examined. The means for accomplishing these ends includes a substrate on which a layer of vanadium dioxide is deposited. The layer is a good reflector at temperatures greater than 67 ^{o}C and reflects only slightly at lower temperatures. The layer is alternately heated by an electrical current and is then allowed to cool to provide the desired optical switching action. This electrically-controlled selectively reflective layer is then used in conjunction with the reference cell and with an array of filters to implement the necessary switching and selection of the components.

Also in 1987, Burough et al. (Burough et al., 1987) disclosed a method and apparatus for detecting a gas. Using this method, an analyzer was invented to detect a specific gas in the atmosphere by infrared absorption. The analyzer has a porous tube surrounding an enclosed column of air, the porous tube having an average pore size to permit gas to diffuse through it quickly. Within the tube are an infrared source, an infrared detector and an optical filter that passes those portions of infrared spectrum most strongly absorbed by the gas. The analyzer has a time base generator to pulse the infrared source and electronic means to measure the maximum and minimum signals from the detector and a meter to indicate the difference between the maximum and minimum as a measure of the amount of gas in the column of air.

In 1988, Thatcher (Thatcher, 1988) disclosed an apnea monitor for preventing sudden infant death syndrome (SIDS). The exhaled breath of the infant is collected in a hood. An infrared source emits infrared energy into the hood. So long as the infant is breathing, the CO_2 in its breath absorbs a portion of the infrared energy in the hood. Should the infant stop breathing, an infrared detector responds to the resulting increase in infrared energy to activate an alarm so as to enable the attendant personnel to take appropriate life saving action.

Also in 1988, Wong (Wong, 1988b) disclosed a compact apparatus for measuring very weak absorption by a gas. It is well known that the measurement of very weak absorption lines can be facilitated by the use of very long transmission path length, which however is difficult to obtain in compact or portable instruments. In Wong's invention, light is made to travel through a limited volume of gas thousands of times. The light is placed in a closed optical path on which it circulates through the gas sample. After a desired number of passes through the gas sample, the light is removed from the closed optical path. Introduction of the light to the closed optical path and removal therefrom is accomplished through the use of a polarizing beamsplitter and a pockels cell located on the closed path. Light is put onto the closed path by the polarizing beamsplitter which imparts a specific polarization. During the first circuit the Pockels cell alters the polarization by 90 degrees thereby preventing the light from escaping back out through the polarizing beamsplitter. After the desired number of circuits, the Pockels cell again alters the polarization by 90 degrees thereby permitting the light to be redirected out of the closed path by the polarizing beamsplitter.

In 1989, Wong (Wong, 1989) disclosed a high speed infrared shutter having no moving parts intended for use at infrared wavelengths. This device includes a thin film of vanadium dioxide deposited on an insulating layer of silicon oxide that has been grown on a thin substrate of semiconductive silicon. The layers are in good thermal contact, but the vanadium dioxide film is electrically insulated from the silicon substrate. The vanadium dioxide film is heated by passing a heating current through the semiconductive substrate, and is cooled by radiation. The resistance of the vanadium dioxide film is sensed continuously, and is used for turning on and off the heating current.

Also in 1989, Schoennauer et al. (Schoennauer et al, 1989) disclosed the working of a resonant piezoelectric chopper for infrared radiation. This chopper is an apparatus for inserting and removing several filters from an optical path and for determining at all times the instantaneous positional relationship of each filter with respect to the optical path. It includes a set of filters mounted on a carrier which is attached to a resonant piezoelectric system. The carrier further includes reference tracks and a sequence of binary indicia extending in the direction of the oscillatory motion. The reference tracks are read optically to produce a series of binary electrical signals which are processed to yield the instantaneous position of the carrier and filters with respect to a fixed optical path.

4.3. Noteworthy Developments in the Period 1991 - 2000

In 1991, Wong (Wong, 1991a) disclosed a spectral ratioing technique for NDIR gas analysis using a differential temperature source. This invention takes advantage of the recent technological advances in infrared components. One component that took advantage of recent advances was the infrared source which was an electrically modulatable infrared micro-source. This micro-source was, in essence, a small thick film resistor pad made out of a special material capable of being heated and cooled at a relatively high rate (up to 100 Hz, typically). Another component that took advantage of then recent technological advance was narrow band-pass filters. Not only could one make an excellent narrow band-pass filter with one center wavelength (CWL) or pass band, filters with two narrow band-passes centered at two spectrally separated wavelengths were then also readily available. By applying a spectral ratioing technique to an instrument for determining the concentration of a particular gas having no moving parts, an extremely compact was designed. The instrument uses as a source of radiation a device that has a radiating element whose temperature is alternated between T_1 and T_2 and whose spectral output approximates that of a blackbody. Radiation from this source is passed through a dual pass-band filter that has two non-overlapping pass bands, one of which is centered at a wavelength where the gas absorbs and the other of which is centered at a wavelength where the sample does not. After passing through this filter, the radiation passes through the sample chamber and is then intercepted by a detector which produces an electrical signal determined by the radiation intercepted. The electrical signal is processed to provide an indication of the concentration of the gas.

A specific use that helped drive the design for new NDIR gas sensors during the 1990s was fire detectors. To understand why this is so, one need only realize that fire is an oxidation process. In order to detect fires, in addition to just detecting smoke from a fire, one must recognize the importance of detecting other byproducts from a fire. Of the three effluent gases invariably generated by common fires, namely CO, CO_2 and H_2O, detection of the first two gases along with smoke is particularly important for realizing a better fire detector than the smoke detector that was commonly available during the 1990s. If designed and constructed correctly, a fire sensor that detects not only smoke, but also CO and CO_2, will likely not only eliminate false alarms, but also will be a faster and a more reliable fire detector.

In other words, in order to realize a better fire sensor than the false-alarm-prone smoke detector for the general public, one needs better gas sensors for the detection of CO and CO_2. Needless to say, these detectors were not limited to the use of any one particular technology. As a matter of fact, any technology that can get the job done will do. Among the most promising technologies that could meet the size, unit cost and performance requirements for detecting CO_2 and CO at that time were NDIR and Electrochemical cell. At least from the standpoint of NDIR, the technology then was a long way from being ready to take on such a challenge. However, it was the goal to realize a better fire sensor than the smoke detector that provided at least some of the driving force behind the design and development of better NDIR gas sensors in the 1990s.

In 1991, Wong (Wong, 1991b) disclosed the design of a simple fire detector that responds more rapidly than the then widely used smoke detectors to some types of fires, such as the flaming ones (like a burning Christmas tree inside a home). This detector senses the increases in the concentration of CO_2 associated with flaming fires. The build-up of CO_2 is sensed by measuring the concomitant increase in the absorption of a beam of radiation whose wavelength is located at a strong absorption band of CO_2. The device is considerably simplified by the use of a window to the sample chamber that is highly permeable to CO_2 but which keeps out particles of dust, smoke, oil, and water. This permits the fire detector to be built of very inexpensive materials.

As it turns out, the detection of CO_2 using NDIR gas sensing technique is relatively easy and straightforward due to the fact that CO_2 has a very strong and specific absorption band at 4.26 μ. Only a very short path length (e.g. 4.0″) sample chamber is needed to detect the gas down to ppm levels. However, in order to adequately detect CO gas down to the same concentration level of a few ppm, a path length of 60.0″ or more would be needed since the 4.65 μ absorption band for its detection is ~10 times weaker than the 4.26 μ absorption band of CO_2. But since the dimensions of a fire sensor utilizing the detection of CO gas has to be quite small, satisfying the very long path length requirement for its detection posed a very difficult challenge for designing its sample chamber. This challenge further provided additional driving forces for the development of such devices and the concomitant gas sensing methodologies.

In 1991, Wong (Wong, 1991c) disclosed a compact gas sample chamber with a very long path length for use in measuring the absorption of radiation as it passes through a gas within the chamber. It is constructed out of a block of material having an extended serpentine passage through it. The walls of the extended passage are coated with a highly reflective material so that the extended passage acts as a light pipe for transmitting the radiation. A number of smaller passages permit gases in the space surrounding the sample chamber to diffuse to the extended passage through which the radiation is conducted.

The sample chamber is made by joining two halves, each of which as a planar face in which an elongated groove is produced. The two halves are molded out of plastic and in quantity the chamber is quite inexpensive.

In 1992, Wong (Wong, 1992a) disclosed a compact low-cost fire detector which responds quickly by detecting an increase in the concentration of CO_2 gas in the ambient air. The detector also calculates the rate of build-up of CO_2. The detector avoids the use of moving parts by employing a differential temperature blackbody source of infrared radiation in conjunction with a dual pass-band filter. One of the pass bands is located at the 4.26 μ absorption band of CO_2 gas and the other pass band is located at 2.20 μ at which none of the atmospheric gases has an absorption band. The latter channel serves as a reference and renders the detector immune to false alarms caused by dust or smoke particles in the air or due to deterioration of certain components. The fire detector makes use of a sample chamber that consists of a serpentine passage in a block of material, the walls of the serpentine passage being highly reflective so as to act as a light pipe.

Apart from the need for a false-alarm resistant fire sensor to replace the smoke detector that was driving the development for better NDIR gas sensing technologies, there appeared in the early 1990s the need for a small, rugged and low-cost CO_2 sensor for building ventilation control to save energy. The so-called Demand Controlled Ventilation (DCV) approach measures CO_2 in different zones in a building to determine the dynamics between the amounts of outside air being introduced to the building versus the actual number of occupants. Compared to the traditional method of sizing outside air delivery to maximum occupancy, a CO_2 sensor can provide important feedback control to regulate the delivery of outside air based on actual occupancy thereby saving energy cost. Thus, another driving force appeared around that time that continued to provide impetus to scientists and engineers to come up with a small, rugged and low-cost CO_2 sensor with no moving parts.

One of the major milestones for the advancement of NDIR gas sensing technology was disclosed by Wong in 1992 (Wong, 1992b). This milestone was a diffusion-type gas sample chamber for use in a gas analyzer made of an elongated hollow tube having an inwardly-facing specularly-reflective surface that permits the tube to function as a light pipe for transmitting radiation from a source to a detector through the sample gas. A number of filtering apertures in the wall of the otherwise non-porous hollow tube permit the sample gas to enter and exit freely under ambient pressure. Particles of smoke and dust of a size greater than 0.1 micron are kept out of the chamber by use of a semi-permeable membrane that spans the apertures in the hollow tube. Condensation of the sample gas components is prevented by heating the sample chamber electrically to a temperature above the dew point of the component of concern. Figure 12 shows a schematic drawing of this so-called "waveguide" sample chamber. As it turns out, this seemingly simple invention provided, for the first time, a simple, small, rugged and extremely low-cost NDIR gas sensor with no moving parts fabricated without the use of optics. For this accomplishment, Wong was bestowed the prestigious *Photonics Circle of Excellence Award* in1993.

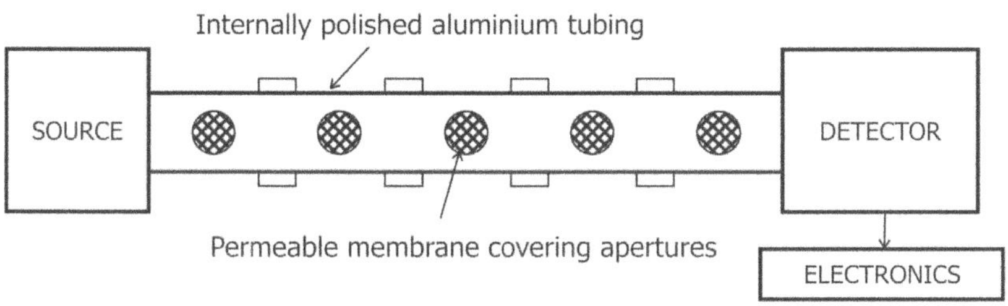

Figure 12. The Award-winning Waveguide Sample Chamber.

In 1993, Wong (Wong, 1993) disclosed a multi-channel gas sample chamber for use in a gas analyzer consisting of an elongated hollow tube having an inwardly-facing specularly-reflective surface that permits the tube to function as a light pipe for conducting radiation from a source to a detector through the sample gas. A number of apertures in the wall of the elongated hollow tube permit the sample gas to enter and exit. Particles of smoke and dust of a size greater than 0.1 micron are kept out of the chamber by use of a semi-permeable membrane that spans the apertures in the hollow tube. Condensation of the sample gas components is prevented by heating the sample chamber electrically to a temperature above the dew point of the component of concern. In one embodiment, at least one detector is spaced around the periphery of the elongated hollow tube adjacent to one end of it. In another embodiment, at least one detector is spaced along the length of the elongated hollow tube.

In 1994, Wong (Wong, 1994a) disclosed a functional testing method for toxic gas sensors. The proper operation of an NDIR toxic gas sensor can be checked by using a non-toxic gas that has an absorption band that overlaps the absorption band used for measuring the toxic gas. The NDIR sensor cannot distinguish which of the two gases is responsible for causing the observed absorption of radiation in the sensor. Since detectors of toxic gases typically operate at low concentration levels, the extent of overlap between the absorption band of the toxic gas and of the non–toxic gas may be relatively small, since a large concentration of the non-toxic gas can be used with no ill effects. The method may be applied to sensors for detecting carbon monoxide (CO) or methane (CH_4) for example.

Also in 1994, Wong (Wong, 1994b) disclosed an improved diffusion-type gas sample chamber which includes an elongated hollow tube closed at one end which has specularly-reflective inwardly facing surfaces. A source of radiation and a detector of radiation are mounted side by side in the open end of the hollow tube. Multiple filtering apertures are formed in the tube, and each aperture is covered by a sheet of a semi-permeable membrane that serves to prevent airborne particles larger than a predetermined size from entering the chamber while not interfering with the free diffusion of the gas to be measured into and out of the chamber. The use of an elongated hollow tube that is closed at one end results in no loss in the efficiency with which the radiation is conducted from the source to the detector while decreasing the external length of the chamber by 50 percent.

Again in 1994, Wong (Wong, 1994c) disclosed an NDIR gas analysis using a spectral ratioing technique. An instrument for determining the concentration of a particular gas that might be present in a sample has no moving parts and is extremely compact and inexpensive. A novel waveguide structure serves both as an optical element and as the sample chamber. As an optical element, the waveguide collects radiation from a blackbody source located at the entrance end of the waveguide and conducts the radiation through the waveguide, concentrating it on two infrared detectors mounted at the opposite end of the waveguide. As a sample chamber, the waveguide causes radiation to undergo multiple reflections that result in the average path length being substantially greater than the physical length of the waveguide. Each of the detectors has its own optical filter, and baffling assures that each detector responds only to radiation which has passed through its filter. One filter defines a spectral passband that coincides with the infrared absorption band of the gas to be measured. The other filter defines a non-absorbing or neutral passband. The electrical signals produced by the detectors are processed to provide a ratio, the value of which is related to the concentration of the particular gas to be detected.

By the middle of the 1990s, the application of Demand Controlled Ventilation (DCV) to control building ventilation in order to save energy started to get the attention of building contractors. But the growth of this practice was disappointingly slow. It did not take long before the reason for this slow growth was identified. The culprit was the output instability over time for most installed NDIR CO_2 sensors. This created a dilemma for building contractors. Unless installed NDIR CO_2 sensors were checked every 6 months to a year, there was no way of telling whether they were still accurate or whether they had drifted. And, if they were found to be inaccurate, they had to be re-calibrated. As a result, the maintenance cost for keeping the installed sensors accurate over time could negate the cost savings promised by DCV. On the other hand, if the installed sensors were left unchecked, their incorrect CO_2 readings could invalidate the benefits of energy savings and, in some cases, could even be wasting energy instead. Thus, there was not any doubt whatsoever that output drift free NDIR CO_2 sensors were needed to overcome this dilemma. Unfortunately, NDIR sensor technology was simply not ready, at that time, to overcome this dilemma.

Fortunately, in 1994, Wong (Wong, 1994d) disclosed a self-calibration methodology that can be applied to an NDIR CO_2 sensor in order to compensate for its output drift over time via software. Some cyclic variables include within each cycle a value that can be determined extrinsically. In such a case, the sensed value may differ from the known value by an amount that is a combination of long-term drift of the sensor and random measuring error. The drift component can be evaluated and eliminated by the following method. Once each cycle, for a number of cycles, the sensor measures the variable at a time when its value should equal the extrinsically-known value. The differences are plotted versus time, and a best-fitting straight line is determined, which indicates the drift. Throughout the next cycle as the variable is continuously sensed, the drift determined from the best-fitting straight line is continuously applied to correct the sensed value.

As it turns out, commercial buildings downtown or near downtown in big cities present one of the optimum circumstances for applying the above described self-calibration methodology to CO_2 sensors used for DCV applications to save energy in order to compensate for inevitable sensor output drifts. One cycle time could be 24 hours or a few days. But the important point to note is that right around 2:00-4:00 AM of every 24 hours is a period where hardly any soul is found inside the building except perhaps one or two guards after the building cleaning crew has long gone home for the night. Also, at this time, few cars are found outside in the streets. Under these circumstances, the CO_2 concentration level inside the building will likely only be between 400 - 450 ppm. Since this condition should presumably repeat itself every 24 hours, this kind of cycles ideal for applying the above described self-calibration methodology for NDIR CO_2 sensors to eliminate output drift over time. In fact, this self-calibration methodology is widely used for NDIR CO_2 sensors in DCV applications and it has a popular name called "Automatic Background Correction," or simply "ABC". Over 90 % of all NDIR CO_2 sensors sold today for DCV applications have the ABC software provision to compensate for sensor output drifts. It is because of the ABC software coming to help to alleviate the problem of output drift for NDIR CO_2 sensors that the practice of DCV for building ventilation control to save energy finally started to move forward in the last two decades.

In 1995, Wong (Wong, 1995a) disclosed a gas sample chamber for use with a source of coherent radiation. A gas sample chamber that is particularly advantageous for use with a semiconductor laser has the form of an elongated hollow tube with a specularly-reflective inside surface, a semiconductor laser located at one end of the tube and a detector located at the opposite end of the tube. In one embodiment, apertures in the wall of the tube permit a gas to enter and leave the sample chamber by free diffusion. In another embodiment, the gas flows into the hollow tube from a pressurized source through a port or is drawn through the tube by a suction pump. In other embodiments, the tube is partitioned into two successive sections by means of a window located inside the tube. The window is transparent to radiation of two different wavelengths that coincide with the absorption bands of two different gases. The semiconductor laser is tuned to these wavelengths successively so that two gas components can be detected and measured simultaneously.

Also in 1995, Wong (Wong, 1995b) disclosed a laser assisted quasi-blackbody radiation source. A radiation was advanced that exhibits an improved efficiency of converting input energy into a quasi-blackbody output beam. A wavelength selective element can be included to transit one or more narrow bands of this radiation for use in spectroscopy. The intensity of the output beam can be varied quickly, thereby decreasing measurement times and enabling synchronous detection techniques to be implemented at reasonable frequencies. A feedback mechanism can be integrated into this device to enable accurate control of the blackbody temperature. This radiation source can be manufactured by conventional integrated circuit manufacturing techniques, thereby producing an inexpensive quasi-blackbody source capable of accurately and repeatably controlled performance.

Also in 1995, Wong (Wong, 1995c, e) disclosed an enhanced pathlength gas sample chamber. The sensitivity of a non-dispersive infrared gas analyzer increases with increasing pathlength, which increases the absorption of the infrared radiation by the gas whose concentration is to be determined. The present invention includes the addition of an obstructing element that further increases the average pathlength. In one embodiment, the obstructing element consists of a highly reflective elongated cylinder located with its axis collinear with the axis of the tubular chamber. In another embodiment, the sample chamber includes a converging section, the smaller end of which is joined to an elongated tubular section of uniform cross section. The improvements are applicable to both pumped and diffusion-type sample chambers.

Also in 1995, Wong (Wong, 1995d) disclosed a miniaturized NDIR gas sensor. A miniaturized NDIR gas sensor is manufactured using semiconductor micromachining techniques from a semiconductor material such as Si or GaAs. The NDIR gas sensor comprises an optical waveguide, a light source at one end of the waveguide, at least one light detector at the end of the waveguide opposite the light source, a diffusion type gas sample chamber formed within the waveguide and interposed in the optical path between the light source and light detector so that the light source and light detector are thermally isolated from the gas sample, and a separate band-pass filter interposed between the light source and each light detector. Because the NDIR sensor is fabricated out of a semiconductor material, the source driver and signal processing electronics may be added directly to the sensor using integrated circuit fabrication techniques. Particles and smoke and dust may be kept out of the sample chamber by application of a gas permeable membrane over apertures in the sample chamber walls.

In 1996, Wong (Wong, 1996) disclosed a diffusion-type gas sample chamber. An apparatus for measuring the concentration of a gas present in a sample chamber by ambient pressure diffusion employs a non-dispersive infrared gas analysis technique. The sample chamber has the form of a tube that is closed at one end, with a source of radiation and a detector mounted side by side at the other end. The inwardly-facing surfaces of the tube are specularly-reflective, whereby the optical length of the sample chamber is twice its physical length. A gas filter cell located in the optical path permits the concentration of an analyte gas to be measured accurately despite the presence in the sample chamber of an interference gas. A small ultrasonic vibrator affixed to the wall of the sample chamber prevents unwanted particles from accumulating on the semi-permeable membranes through which ambient gases diffuse into and out of the sample chamber.

In 1997, Wong (Wong, 1997a) disclosed a false alarm resistant fire detector with improved performance. Such a fire detector has a greatly reduced frequency of generating false alarms. An AND gate is responsive to outputs from first and second fire detector modules that are responsive to the detection of first and second characteristics of a fire, respectively, to signal the detection of a fire if both of these characteristics have been detected. At least one override path is included to enable the occurrence of a particular type of fire to be signaled that would not otherwise be signaled by the output of said AND gate.

Also in 1997, Wong (Wong, 1997b) disclosed a passive infrared analysis gas sensor. This sensor uses an ambient temperature source and the space between the detector assembly and the source as the sample chamber. The gas detector includes an infrared detector assembly for producing a first output, a second output, and a third output, the first output being indicative of the radiation received by the detector assembly at a first non-neutral spectral band which is absorbable by a preselected gas to be detected, the second output being indicative of the radiation received by the detector assembly at a first neutral spectral band from the passive infrared source, and the third output being indicative of the radiation received by the detector assembly at a second neutral spectral band from the passive infrared source. Signal processing means are included for manipulating the three outputs to determine the concentration of the gas being monitored. By adding additional detectors to the detector assembly which can detect radiation at spectral bands characteristic of additional gases, the infrared gas detector can be used to monitor the concentration of a plurality of gases.

Again in 1997, Wong (Wong, 1997c) disclosed a practical and improved fire detector which combines an NDIR CO_2 gas detector with a photoelectric smoke detector to minimize false alarms by logic means that can be integrated into a single chip that can have an ASIC section and a microprocessor section. The NDIR CO_2 detector can be single or dual channel. The NDIR CO_2 gas detector and the photoelectric smoke detector can be separate or combined in a single device in which they are optically isolated by a light-tight barrier but still use a common light source. Also the CO_2 and smoke detectors can be combined on a single substrate within a common housing.

In 1998, Wong (Wong, 1998) disclosed passive and active infrared analysis gas sensors and applicable multi-channel detector assemblies. In this invention, multi-channel infrared detector assemblies for use in the detection and monitoring of gas concentrations are provided. The detector assemblies include a detector housing having a port for receiving infrared radiation through; a substrate mounted within the detector housing, the substrate having three apertures therein to transmit radiation entering the detector assembly through; a first, a second and a third thermopile detector fabricated on the bottom side of the substrate with the hot junctions of each thermopile detector positioned over one of the apertures in the substrate so as to receive radiation transmitted through the aperture, and the cold junctions of each thermopile detector positioned over the substrate; a first interference band-pass filter mounted on the top side of the substrate so that the first filter covers the aperture above the first detector and the first filter is interposed between the port and the first detector. The first interference band-pass filter is designed to pass incident radiation at a first spectral band. A second interference filter is mounted on the top side of the substrate so that the second filter covers the aperture above the second detector and the second filter is interposed between the port and the second detector. The second interference band-pass filter is designed to pass radiation at a second spectral band. A third interference band-pass filter is mounted on the top side of the substrate so that the third filter covers the aperture above the third detector and the third filter is interposed between the port and the third detector. The third interference band-pass filter is designed to pass radiation at a third spectral band. The disclosed infrared detector assembly can be used in traditional NDIR gas sensors having an active source or in passive infrared analysis gas sensors which use a passive infrared temperature source and the space between the detector assembly and the source as the sample chamber.

4.4. Noteworthy Developments in the Period 2001 – 2010

In 2001, Lampert et al. (Lampert et al., 2001) disclosed a dynamic infrared sensor for automotive pre-vaporized fueling control for use in fueling an internal combustion engine. The infrared sensor senses the hydrocarbon content of the vaporized fuel which information is used by the engine control module to control engine fueling to minimize emissions. The infrared source is operated to maintain in the pre-vaporized fuel below its lower explosion limit and the thermopile detector electronics is synchronized with the light pulse frequency to develop fast signal response suitable for fueling control.

In 2007, Wong (Wong, 2007a) disclosed a passive NDIR gas sensor fire detector. A fire detector and method for detecting fires uses a passive NDIR gas sensor that generates a detector signal based upon either a 15 μ absorption band of CO_2 or a 6.27 μ absorption band of H_2O and a signal processor which receives the detector signal and generates an alarm signal when a preselected criterion is met. The NDIR sensor relies upon a passive infrared source (such as a very thin film of black plastic material) with a high emissivity (e.g. approximately 1.0), a sample chamber with a low emissivity (e.g. approximately 0.03) thermally decoupled from the passive infrared source and an infrared detector located in a detector assembly thermally coupled to the sample chamber while a heat exchanger is thermally coupled to the sample chamber. Gas entering the sample chamber passes through the heat exchanger and both it and the infrared detector are at a lower temperature than that of the passive infrared source at an onset of a fire while the difference in temperature between the passive infrared source and the detector increases during the onset of a fire. The sample chamber and the heat exchanger can be formed integrally from aluminum and their internal surfaces may be coated with a hydrophobic membrane.

Also in 2007, Wong (Wong, 2007b) disclosed a passive NDIR Carbon Dioxide sensor fire detector. Such a fire detector with a passive CO_2 sensor generates a detector signal based upon a 15 μ absorption band of CO_2 and a signal processor generates an alarm signal when a preselected criterion is met. The passive CO_2 sensor has a passive infrared source with a source temperature, a waveguide sample chamber with a gas medium temperature (which is less than the source temperature), a heat exchanger (that reduces the temperature of CO_2 gas as it passes into the waveguide sample chamber and may be thermally couple to the sample chamber) and an infrared detector assembly with a detector temperature (which is less than the source temperature). An exterior surface of the passive infrared source has a high emissivity while the detector assembly and the sample chamber (which can be thermally coupled to each other but not the passive infrared source) has a low emissivity.

Also in 2007, Wong (Wong, 2007c) disclosed an ultra low power NDIR Carbon Dioxide sensor fire detector with method for generating an alarm signal in response to a fire. The NDIR CO_2 sensor is used to generate a detector signal based upon a 15 μ absorption band of CO_2 and generates an alarm signal when a signal processor receives the detector signal and a preselected criterion is met that is indicative of the onset of a fire based upon an analysis of the detector signal using a detection algorithm that relies upon a trending pattern of the detector signal such as recognizing a substantial drop in the detector signal strength. The NDIR CO_2 sensor can also generate a reference detector signal based upon a 9.0 μ neutral band with a FWHM = 0.5 μ while the signal processor utilizes a detection algorithm that is based upon a synchronized output signal representative of CO_2 concentration to generate an alarm signal when a preselected criterion indicative of the onset of a fire is met.

Also in 2007, Tse et al. (Tse et al., 2007) invented a low-cost noise-immune wireless methodology for Demand Controlled Ventilation (DCV) applications. This is a method for reducing the cost of demand controlled ventilation (DCV) applications and deploys wireless CO_2 sensors inclusive of relative humidity and temperature throughout an indoor space without the need for wiring while utilizing the license-free 902-908 MHz ISM band by use of a Sequential Transmission Asynchronous Reception Systems ("STARS") that is software controlled.

Also in 2007, Wong (Wong, 2007d) disclosed a method for detecting a gas species using a super tube waveguide. This is a method for detecting the concentration of one or more gas species by using infrared radiation emitted from one or more sources into a sample cell which is a hollow waveguide with multiple bends collectively greater then 180 degrees in three dimensions, the infrared radiation being quasi-focused into a beam with an angle of incidence $0 - 10^{\circ}$ relative to a longitudinal axis of a first linear segment of the sample cell proximate the source, then detecting two or more signals in which one of the signals is used to compensate for water vapor. The methods can detect gas concentrations down to 1 ppm or less.

In 2008, Wong (Wong, 2008a) disclosed a simple multi-channel NDIR gas sensor. Concentration of N gas species can be detected from a single beam NDIR gas sensor having a differential infrared source and an (N+1)-pass-band filter (having a neutral pass-band and N absorption pass-bands for N gases) mounted at a single infrared detector. By driving the infrared source into emitting at N distinct temperatures, whose radiation outputs are characterized by N corresponding Planck curves which are dependent only upon the respective source temperatures and which link a Spectral Radiant Emittance (M_λ) with wavelength, the N detector outputs at the single infrared detector are then measured and the concentrations of N different gas species detected, each of the N gas species having its own unique infrared absorption pass-band. This is done by (a) setting up N causality relationship equations liking outputs of the detector respectively for N different source temperatures and a set of relevant parameters of the sensor components, (b) determining the values of all of the parameters for the N equations utilizing appropriate boundary conditions except the N concentrations for the respective N gas species, and (c) solving for the N concentrations with the measured N detector outputs, there being N equations and N unknowns, where N is an integer of 2 or more.

Again in 2008, Wong et al. (Wong et al, 2008b) disclosed an ultra low cost NDIR gas sensor. The concentration of a gas species is detected by using a single beam NDIR gas sensor in which an infrared source element is driven at two different temperatures, a feed back loop senses an operation voltage of the source, a differential gain amplifier creates a high cycle amplified output during a high cycle and a low cycle amplified output during a low cycle while a controller synchronizes the source driver so that a signal processor can determine the gas concentration through use of the high cycle amplified output and the low cycle amplified output. The infrared source can be a non-genuine blackbody source such as an incandescent miniature light bulb when the sample chamber is a thermally insulated aluminum tube that is maintained at a preselected temperature greater than ambient so that the glass envelope of the bulb is maintained at an equilibrium temperature of ~$30^{o} \pm 2\,^{o}C$ during its low cycle operation state.

Chapter 5

NDIR Gas Measurement Today

5.1. A Short Summary of What Led Up to Today

As noted in Chapter 1, Non-dispersive infrared (NDIR) gas sensors have long been considered one of the best methods for gas measurement since the 1960s. Other gas detection methodologies include technologies such as electrochemical cells, catalytic (platinum bead) sensors, photo-ionization detectors, flame-ionization detectors, etc., which are generally referred to as "interactive" types of gas detectors. Interactive sensor technologies tend to be subject to interference from other gases and environmental conditions and generally degrade over time resulting in a limited useful life. A superior gas measurement technique, which only became fully developed within the past two decades, is tunable diode laser absorption spectroscopy (TDLAS) methodology. This technique uses the output of a laser as a tunable coherent radiation source; however, TDLAS sensors have not been widely deployed to date due to the limited availability of laser wavelengths that cover the absorption bands of many important gases, and the necessity of having to resort to the use of very expensive quantum cascade type lasers to cover such absorption bands. Other more advanced optical gas sensors utilizing infrared emitting diodes (ireds) fabricated out of III-V ternary semiconductor material, such as $InAs_{1-X}Sb_X$, are often deployed in special applications where the sensor size, its speed of response (one millisecond or less) and ultra low power consumption (< 1 mW) are of critical importance. New gas sensors that are coming along utilizing Nano-technology clearly have to be seriously considered for use in the future. But, unlike the other well-established and trusted gas sensing technologies, Nano-technology type gas sensors will have to be tested for years before they will be accepted as reliable. With this background in mind, until the end of the last decade, NDIR gas sensors, while not perfect by any means, were still considered highly gas specific, sensitive, fast responding, relatively stable over time, rugged, reliable and easy to use and maintain, and available for a reasonable cost.

But, despite over six decades of incessant technology development for NDIR gas sensors, there was still one conspicuous and very important sensor performance deficiency yet to be overcome. This is sensor output instability or drift over time. It is a deficiency that has limited more widespread adoption and use of NDIR gas sensors. For example, without an output stable CO_2 controller, implementation and practice of demand controlled ventilation (DCV) strategy in office and commercial buildings to save energy is cost-wise very difficult to justify due to the expensive recalibration routine needed every 6 months to a year to maintain the accuracy of such sensors. This was the case until the self-calibration methodology of sensor output correction software was developed in the 1990s (see Section 4.3 of Chapter 4 above). Also, some companies selling NDIR CO_2 controllers in the field claimed to have successfully utilized automatic source adjustment to take care of aging effects in order to combat subsequent sensor output drifts over time. But this non-publicly disclosed method has never been proven consistently to be valid or true. Thus, the only workable solution for temporarily solving this problem, so as not to stymie the HVAC&R industry's implementation of DCV strategy to save energy, was through the use of sensor output correction software.

Sensor output correct software was disclosed in a USPTO patent issued to Wong in 1994 (Wong, 1994d). Such methodology can be applied to an NDIR CO_2 sensor in order to compensate for its output drifts over time via software. But this methodology is valid only under some very special circumstances, and it is not applicable to the detection of water vapor. Fortunately, using this methodology the practice of DCV to office and commercial buildings located in some special areas, such as downtown or near downtown, can be effectively implemented. A CO_2 controller equipped with such sensor output correction software is commonly referred to as having the "Automatic Background Correction" or simply the "ABC" feature installed. Today close to 100 % of all CO_2 controllers used in the application of DCV to office or commercial buildings are equipped with this "ABC" output correction software. It is by no means an over statement to say that without the "ABC" software methodology the implementation status of DCV strategy to office and commercial buildings would not have been as robust as what it is today. Still, it is estimated that only 3 % of all the eligible buildings in the world today, including several million government buildings in the United States, have implemented the DCV strategy to save energy.

5.2. The Turning Point: Understanding the Physics of NDIR Gas Measurement

Until the end of 2010, the problem of output drift for NDIR gas sensors remained a major and seemingly never-ending challenge to researchers and engineers in the field of NDIR gas sensors. A breakthrough finally came when Wong and Schell published a research paper on the subject in 2011 entitled "Zero Drift NDIR Gas Sensors" in *Sensor Review* (Wong et al., 2011c). In this published work, the authors described and explained in detail the Physics behind NDIR gas measurement methodology.

In order to properly describe and explain the Physics behind NDIR gas measurement methodology so as to truly understand it, one has to start from the very beginning. The NDIR gas measurement technique first conceived many decades ago took advantage of the fact that quite a large number of common and important gases have their individual and characteristic vibration/rotational absorption spectra in the infrared spectral region between ~2 μ and ~20 μ with narrow and non-overlapping lines or bands. These absorption bands, unique for different individual gases, are commonly referred to as their infrared signatures. For the detection of a particular gas, the NDIR gas measurement technique first sets up a band of infrared radiation that best fits the signature of the target gas. Because of absorption, the amount of target gas present in this infrared radiation band will proportionately diminish its strength or intensity. By monitoring the lost strength or intensity of this band with a detector means, the amount of target gas present is determined. Thus, at least in principle, NDIR gas measurement technique is indeed rather simplistic.

In order to implement NDIR gas measurement technique in some form of a device according to the above methodology, there are five items of information or objects that one needs. They are: 1) a broadband infrared source; 2) a detailed signature of the target gas; 3) a sample chamber; 4) a selector of infrared radiation band from the source; and 5) an infrared detector and its spectral response characteristic. Once these five items of information or objects are identified, an NDIR gas sensor for any target gas can be designed and constructed.

In the design of NDIR gas sensors, almost all infrared sources without exception are graybodies, so-called because the emissivity of these sources is always less than unity. However, Planck's radiation law comprising a family of spectral radiation curves for sources operating at diverse temperatures represents equally well both graybody (emissivity < 1) and blackbody (emissivity = 1) sources. Figure 7 (see Chapter 1) shows the radiant emittance excitance curves of a blackbody source at temperatures between 100 and 1,000 ^{o}K. One can see from Figure 7 that within the infrared spectral region of ~2 μ to ~20 μ, where the signatures of many important gasses are located, one can always find and operate a graybody source that will meet the design needs of an NDIR gas sensor.

The detailed signature of a target gas useful for the design of NDIR gas sensors is its infrared absorption band. Over the decades, the absorption spectra of many gasses have been extensively recorded, analyzed and tabulated. Today the absorption spectra of many common gases can be found in a number of sources (Infrared Analysis, Inc., 1995). Their signatures are normally available in the form of an absorbance versus wavelength curve. An example is shown in Figure 4 (see Chapter 1) for CO_2 gas. Figure 4 shows the infrared absorption band of CO_2 at 4.259 μ with its P, Q and R branches. In this figure the absorbance value of CO_2 is plotted versus wavelength. The absorbance value of the gas is linked directly to the absorption coefficient "k" from the Beer-Lambert Gas Absorption Law (see Section 1.3.4 of Chapter 1 above).

A sample chamber is simply a space to confine the target gas so that it is exposed to the radiation coming from the infrared source before the radiation reaches the detector. This will allow the detector to determine the lost radiation coming from the source because of the presence of the target gas in the sample chamber. Over the years many different designs and forms of sample chambers have been advanced, tried and improved. Surprisingly, this is one area in which great progress has been made in research and development of NDIR gas sensors over the years.

Interesting enough, it was the selector of infrared radiation band from a broadband infrared source that gave birth to this methodology for gas measurement and also its somewhat unusual name. As early as over three centuries ago, prisms and diffraction gratings were considered "dispersive" elements in optics that had the ability to spatially separate light of different wavelengths sent past them. As told earlier in this book (see Chapter 2), the concept of a small, compact and light-weight optical element capable of spectrally separating electromagnetic radiation of different wavelengths was first conceived in the 1930s as part of classified work in the development of night vision devices by the U.S. military. Regular dispersive elements such as prisms and diffraction gratings were considered to be too bulky and heavy for use and the fact that dispersive elements separate wavelengths of light via spatial deflections was another of their undesirable features for design. The final outcome was the invention by the U.S. military of a non-dispersive optical element which was the multi-layer interference filter capable of isolating a narrow spectral band of light and rejecting all others without deflecting the beam of broadband radiation incident at it. The U.S. Defense Department declassified the use of this so-called narrow band-pass interference filter to the general public in 1948 but it was not until the middle of the 1960s before actual working devices were available for use by gas sensor manufacturers. The non-dispersive infrared (NDIR) gas measurement technique got its name simply because it uses a narrow band-pass interference filter, a non-dispersive element, as the selector of infrared radiation band from a broadband source.

The fifth and last item of information or object that one needs in order to implement the NDIR gas measurement technique into a practical gas sensor is the detector. The detector must possess a spectral response characteristic that covers adequately the spectral signature of the target gas. Of the eight most common types of infrared detectors commercially available today (which are shown in Table 1), only two types-- Pyroelectric and Thermopile detectors--have spectral response that is "White," meaning that the spectral response covers all wavelengths. Both Pyroelectric and Thermopile detectors are thermal detectors. Although thermal detectors are typically less sensitive and slower than the other types, they are also the most inexpensive. As can be seen from Table 1, some detector types can have detection sensitivity (as measured by their D^*) orders of magnitude higher than thermal detectors. However, their unit cost can also be orders of magnitude higher, which makes the role played by the detector in the design of NDIR gas sensors critical. For gases having strong signatures, when the required detection sensitivity and response time are modest, such as, for example, CO_2, inexpensive thermal detectors are quite adequate to get the job done. However, for a gas having a very weak signature, when the required detection sensitivity is high, such as, for example, CO, thermal detectors might still work if one is willing to sacrifice sensor size by going to a very long path length sample chamber (~60 inches or more). Otherwise other detector types as listed in Table 1 will have to be deployed to meet the requirements at the expense of skyrocketing unit sensor cost.

Table 1. Commercial Infrared Detectors.

Detector	Type	Oper T °K	λ_{cutoff} (μ)	D* ($cmHz^{0.5}W^{-1}$)	Unit cost, $
Pyroelectric	Thermal	Room	"white"	10^8	10
Thermopile	Thermal	Room	"white"	10^8	10
PbS	PC	Room	3.0	10^9	50
PbS *cooled*	PC	193	3.5	10^{10}	150
PbSe	PC	Room	4.0	10^9	75
PbSe *cooled*	PC	193	4.6	10^{10}	200
InAs	PV	Room	3.5	10^9	50
InAs *cooled*	PV	193	3.2	10^{11}	100
InSb	PC, PV	Room	6.5	10^9	50
InSb *cooled*	PC, PV	77	5.5	10^{11}	200
HgCdTe	PC, PV	Room	3-6	10^9	200
HgCdTe *cooled*	PC, PV	77	3-6	10^{11}	500-1,000
Extrinsic Si	PC	77	>10	10^{10}	500

The first people back in the early 1960s who tried to come up with a working gas sensor without much thought by simply putting together the five items of information or objects listed above to implement the NDIR gas measurement technique probably were quite satisfied but also rather disappointed. Today their setup is generally referred to as the Single Beam Configuration Design (see Section 1.3.4 of Chapter 1). The reason we are guessing today why they were quite satisfied is that the NDIR gas sensor they put together actually worked. The reason why they were rather disappointed is that the output of the sensor was affected by a number of factors like dust particles and environmental temperature. In particular, the output simply did not remain constant over time. Today we fully understand the reasons why the NDIR gas sensors they built behaved that way. But it did not take the scientists and researchers long at that time to also find out the reasons why the performance of their NDIR gas sensors they built were so poor. They soon discovered that an effective way to eliminate or reduce the error-causing factors inherent with the Single Beam sensor design they came up with was to include a second channel which operates at a neutral wavelength just off the absorption one but in no way coincident with it. This so-called Reference channel is equally affected by all non-gas phenomena as the Signal channel, so by taking the ratio of the signal outputs from the two channels, namely the Signal and the Reference, one can ideally eliminate in theory the effect of all the error-causing factors inevitably present in the Single Beam sensor design. Only the presence of the target gas in the sample chamber should affect the ratio. Today we call this configuration the Double Beam NDIR gas sensor design (see Section 1.3.4 of Chapter 1).

This so-called Double Beam sensor design has been in use in this industry for over four decades. But until the end of 2010, the problem of output drifts for NDIR gas sensors over time still persisted and nobody seemed to understand why or simply just did not care. Finally, in their published paper entitled "Zero drift NDIR gas sensors" in *Sensor Review* (Wong et al., 2011c), Wong and Schell explained for the first time the reason why. Taking the NDIR CO_2 gas sensor as an example, the authors superimposed the spectral locations for the Signal channel 3 and Reference channel 4 narrow band-pass spectral filters, respectively at 4.26 μ and 3.91 μ on Planck's Blackbody radiant emittance excitance curves of Figure 7 (see Chapter 1). One can see from Figure 7 that if the temperature of the source changes, thereby shifting the Planck's spectral radiation excitance curve, the relative magnitudes of the radiation traversing the filters at their respective spectral locations will change and so will the ratio of their respective detector outputs. This is precisely the reason why the outputs of all NDIR gas sensors today that utilize this so-called Double Beam methodology drift over time as the infrared source ages, resulting in a change of its spectral output intensities at the Signal and Reference filter wavelengths. Thus, the sensor output expressed as the ratio of the Signal channel detector output over the same for the Reference channel can never be held constant over time. Against the calibration curve for the sensor, the change of this ratio value is equivalent to a change or drift in the sensor output as time goes forward.

5.3. Reaping the Benefits: Zero Drift NDIR Gas Sensors

From the discussion above, using a spectral location where there is no infrared absorption by the gas of interest for establishing a reference channel will not prevent the output of a Double Beam NDIR gas sensor from drifting over time. In order to design a Double Beam NDIR gas sensor whose output remains substantially drift free in the course of time, one must make sure that there are no spectral content discrepancies delivered at any time to both the Signal and the Reference channels. Only in this way will the ratio of the Signal channel detector output over the Reference channel detector output stay substantially invariant even as the infrared source ages to thereby change its spectral output or as the temperature of the surroundings of the sensor changes unpredictably over time.

After understanding the reason why the outputs of conventional Double Beam NDIR gas sensors invariably drift over time, Wong advanced a so-called Absorption Biased (AB) Double Beam sensor design methodology in order to eliminate or significantly suppress this problem (Wong, 2011b). Wong achieved this feat in his methodology by first applying the same spectral band-pass filter to both the Signal and the Reference channels. An absorption bias is then additionally applied to the Signal channel by making the sample chamber path length associated with it longer than that associated with the Reference channel. Figure 9 (see Chapter 1) shows the optical component layout for the design of an NDIR gas sensor following Wong's Absorption Biased Double Beam methodology. As shown in Figure 9, both the Signal channel detector 5 and the Reference channel detector 6 are entrapped with 100 % nitrogen 7 and have the same narrow band-pass filter 8 which is used to detect the gas of interest in the sample chamber 9. As an example, the filter designed to be used for the detection of CO_2 gas will have a CWL = 4.26 μ and a FWHM = 0.14 μ. Notice that both detectors 5 and 6 are thermally connected to the entire sensor body 10 through their respective waveguides 11 and 12 and consequently they always share the same thermal platform with each other. In other words, the entire sensor body 10, which is in essence a composite of aluminum parts comprising the infrared source mount 13, sample chamber 9 and the waveguides 11 and 12, respectively, for the Signal and Reference channels, provides an excellent common thermal platform for detectors 5 and 6.

As shown in Figure 9, the sample chamber pathlength L_R, 14, associated with the Reference channel is approximately one-half of the sample chamber pathlength L_S, 15, associated with the Signal channel. A common infrared source 16 is used to illuminate both the Signal and the Reference channels. The output of detector 5 for the Signal channel is always less than that of the detector 6 for the Reference channel irrespective of whether or not there is any amount of the gas of interest in the sample chamber 9. The respective detector outputs can be determined with the use of the Beer-Lambert Absorption Law (see Section 1.3.4 of Chapter 1) for the particular gas of interest and the designed characteristics of the narrow band-pass filter 8 together with the physical dimensions for L_R 14 and L_S 15.

Figure 13 shows the graph 17 depicting the output of the Reference channel detector 6 as a function of CO_2 concentration in the sample chamber 9. Graph 18 of Figure 13 shows the output of the Signal channel detector 5 as a function of CO_2 concentration in the sample chamber 9. Notice that the detector output of the Signal channel detector 5 is always less than that of the Reference channel detector 6. The reason is that the sample pathlength L_R of the Reference channel is shorter than the Signal channel sample pathlength L_S. Therefore, whenever there is CO_2 gas present in the sample chamber 9, Signal channel detector 5 will experience more absorption by the CO_2 gas than the corresponding Reference channel detector due to its longer associated pathlength L_S. Graph 19 of Figure 13 shows the ratio of the Signal channel detector output over that of the Reference channel detector as a function of CO_2 concentration in the sample chamber 9. Graph 19 of Figure 13 is the de facto calibration curve for the Absorption Biased (AB) NDIR gas sensor.

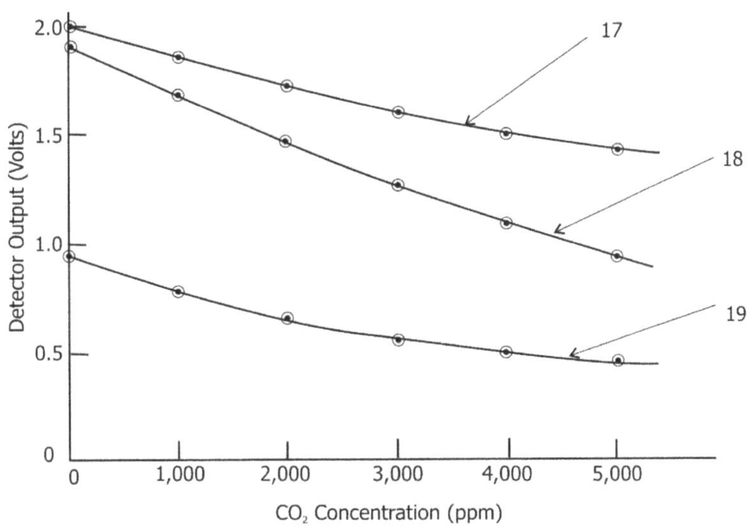

Figure 13. Outputs for Signal and Reference Channel Detectors as a Function of CO_2 Concentration and the de facto Calibration Curve for an Absorption Biased (AB) Designed NDIR Gas Sensor.

If this Absorption Biased (AB) methodology for the design of NDIR gas sensors really works, the output drift of this type of designed sensors will be eliminated or at least significantly suppressed over time. In trying to validate this Absorption Biased design technique, scientists at Oak Ridge National Laboratories were approached to evaluate the theoretical basis for this approach as a methodology for eliminating output drifts in infrared gas sensors. In a letter dated September 30, 2009, Bruce Warmack, Distinguished Research Staff and Som Shrestha, R&D staff, confirmed that the "Absorption Biased (AB) or path length correlation design eliminates sensor drift caused by source variations" and "addresses the central weakness in existing NDIR sensor calibration stability" (Warmack et al., 2009).

While a positive theoretical evaluation by experts in the field on the Absorption Biased (AB) methodology for designing NDIR gas sensors is a good thing, the final determination of its capability must be confirmed by actual available test data. An experiment was therefore designed to test the efficacy of this Absorption Biased design approach. A CO_2 sensor prototype was designed to incorporate the Absorption Biased (AB) design as described above. Since the spectral output drift of an infrared source is proportional to its changing operating (Blackbody) temperature over time, an experiment was devised where the voltage supplied to an incandescent light bulb as the infrared source could be varied in order to simulate such source output drifts, both short and long term. In the experiment, the voltage to the source was modulated in three different steps in the following sequence: 4.7, 4.5, 4.2, 4.5, and 4.7 V. This represents a 20 % modulation of the input power to the source that affects its power output typically with 60-70 % efficiency. In a conventional CO_2 sensor a modulation of less than 1 % in the source power output can cause a significant modulation of the signal.

In Figure 14, the top of the chart defines the voltage used to drive the incandescent source of the sensor. Line 1 shows the variation in temperature that occurred during the 4 hr. and 20 min. test. Since this test was not designed to demonstrate sensor performance versus ambient temperature, Line 1 was shown only for the sake of completeness. Line 2 shows the output of the Signal channel detector during the test. Line 3 shows the output of the Reference channel detector that utilizes an identical infrared filter to the Signal detector but has a much shorter path length as is required by the Absorption Biased (AB) design approach. Line 4 shows the ratio of the Signal detector output over the Reference detector output. As can be seen, the change in a 20 % variation in the drive voltage for the source (which is directly related to the temperature drift of the Blackbody characteristics for the source) resulted in a flat line output, which corresponds to the CO_2 level held at a constant level in the chamber during the test. These experimental data provide conclusive verification that the Absorption Biased design methodology for NDIR gas sensors is capable of compensating for a dramatic change in source output while still reading the correct gas concentration. In other words, for the first time, the problem of NDIR gas sensor output drifts over time can now be eliminated or significantly suppressed with the adoption of the Absorption Biased design methodology.

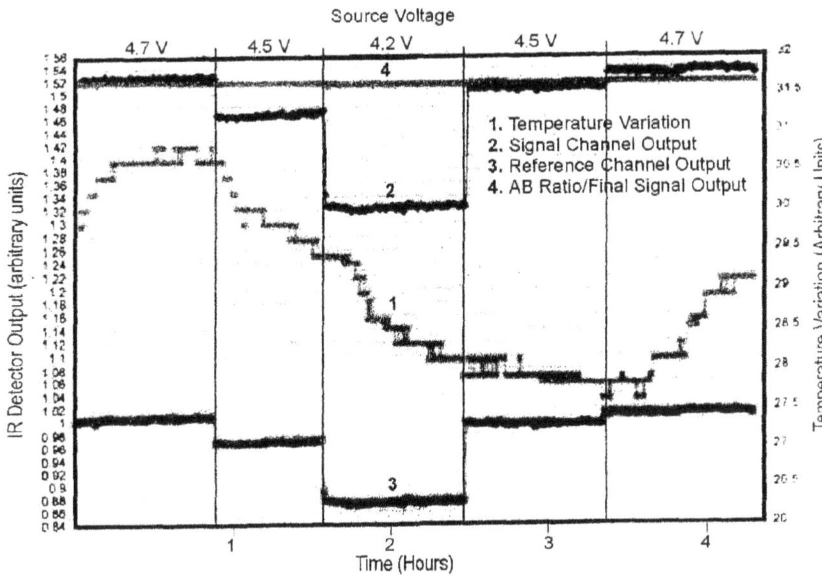

Figure 14. Demonstration of output stability versus simulated source drifts for an Absorption Biased (AB) designed NDIR gas sensor.

5.4. Nothing can be Perfect: Recalibration without Standard Gases

During the 1960s when workers in the field of developing NDIR gas sensors found out that the Single Beam designed sensors had a lot of inherent error-causing factors attributing to their extremely poor performances, they quickly developed the Double Beam designed sensors. By including a second or Reference channel which operates at a neutral or non-absorbing wavelength of the target gas and by processing the ratio of the first or Signal channel detector output over the second or Reference channel detector output as a function of the target gas concentration in the sample chamber, most of the error-causing factors they observed in the Single Beam sensors were eliminated or significantly suppressed. But even with this improvement, one prominent deficiency still persisted until the end of the last decade, which was of course sensor output instability or drift over time.

There is hardly any doubt that NDIR gas sensors today can be designed using the Absorption Biased (AB) methodology to have negligible output drifts over time, or at least significantly lower output drifts over time than have been previously obtained. The AB design methodology is aimed specifically at eliminating the effect of any spectral output change of the source on the sensor calibration curve as it ages thereby significantly suppress or completely eliminate the sensor output drifts over time. This capability of the AB design methodology has been successfully and unequivocally proven in the laboratory by bona fide experimental data. However, while the AB designed NDIR gas sensor can indeed neutralize the effect of any spectral output changes of the source over time, it might not be effective at all in dealing with changes of other sensor component characteristics. Since we are dealing with the ratio of the Signal channel output over the Reference channel output, any spectral or non-spectral imbalance of the two channel outputs will affect the sensor calibration curve and consequently its output. In other words, if the characteristics of a sensor component belonging to the Signal channel changes differently over time from the characteristics of a counter sensor component belonging to the Reference channel, then an imbalance will exist between the two channels. Such an imbalance will cause the ratio of the two channel outputs to change thereby leading to a sensor output change or drift over time.

One example has to do with the responsivity of the two detectors, namely the Signal channel detector and the Reference channel detector. Assuming that over time, albeit maybe a very long time, the responsivity of the Signal channel detector degrades at the spectral location of the band-pass filter a little faster than that of the Reference channel detector, then the ratio of the two channels will change leading to an output drift for the sensor. Unfortunately, even the AB designed NDIR gas sensor will not be able to eliminate or suppress this output drift.

Another example has to do with the infrared source itself, for example, the incandescent light bulb. So far we have assumed that as the source ages over time, only its spectral output changes, but that change of course will be adequately compensated for by the AB design methodology. However it is also possible that over time, albeit again maybe over a very long time, the aging of the source will be reflected not only in its spectral output change, but also in the sagging of its filament due to the continuous evaporation of the tungsten. Unfortunately, the sagging of the source's filament will change its physical radiation pattern and that will inevitably create a non-spectral imbalance between the Signal and the Reference channels because they are physically separated. Again, this imbalance cannot be remedied by the AB design methodology and it will also lead to sensor output drift over time.

As a third example, over time, again albeit a very long time, the reflectivity of the Signal channel waveguide could degrade differently from that of the Reference waveguide due to the prolonged exposure to very fine particles and water vapor despite the protection of air filters. Again, this will result in a non-spectral imbalance that leads to sensor output drift which is incapable of being compensated for by the AB design methodology.

In conclusion, we have just shown in the above discussion that nothing can really be perfect in this world and that includes, of course, NDIR AB designed gas sensors. In other words, despite the fact that AB designed NDIR gas sensors will generally remain drift free for quite a long time, maybe a few years or even longer, these sensors will not stay drift free within certain specifications (such as ± 100 ppm) forever. At some point in time, they must be checked to see if their accuracy is still within their specified limits. Facing this reality that nothing can be perfect in this world, the advent of the AB design methodology for NDIR gas sensors might only be a step towards perfection. Moreover, the fact of the matter is that people have experienced gas sensor output instability for so long that even if output stable sensors really come along, nobody will believe it. Until such time that stable gas sensors become widely available and users begin to believe that their performance is trustworthy, the real need today must be viewed in a completely different perspective. Thus, the real need today is for a fast, inexpensive and simple methodology that can easily check the accuracy of gas sensors and, more importantly, just as easily and simply, recalibrate them when they are found to be inaccurate, all at a cost that is not prohibitive.

Recognizing the fact that it will probably take quite some time before the full impact of output drift free AB designed NDIR gas sensors is felt by the general public, and, in particular, by the HVAC&R industry, Wong in 2011 (Wong, 2011d; Wong et al., 2012a) disclosed a novel re-calibration methodology for AB designed NDIR gas sensors that is simple, easy and does not need to use standard gases. His thorough understanding of the Physics behind the NDIR gas measurement methodology over the years led him to a special formulation for the AB designed NDIR gas sensor calibration curve that identifies two distinct functional domains. One domain is identified to involve only the Physics measurement constituent of the methodology and therefore it should always remain invariant with respect to the sensor's output performance. The other domain is identified to be the sensor components constituent and it is responsible for the sensor's variant output behavior over time. Furthermore, Wong introduced the concept of "Gamma" which is the ratio of the Signal detector output over the Reference detector output when there is no target gas present in the sample chamber. Since Gamma has nothing to do with the measurement Physics of the target gas, it represents only the variant domain characteristics for the sensor components. By normalizing with Gamma the de facto calibration curve for an AB designed NDIR gas sensor (see Graph 19 in Figure 13 for CO_2 gas), Wong is able to combine the two functional domains together to form the resultant calibration curve for the sensor (see Graph 20 in Figure 15 for CO_2 gas). Notice that Graph 20 in Figure 15 always starts out with the value unity when there is no target gas present in the sample chamber.

This new formulation of the calibration curve for an AB designed NDIR gas sensor as shown by Graph 20 for the CO_2 gas in Figure 15 enables one to separate the invariant Physics constituent of the NDIR gas measurement principle from the other inevitably changing components constituent of the sensor over time. In other words, any changes in the calibration curve of an AB designed NDIR gas sensor will only be reflected in the changing value of Gamma over time. It will not be reflected in the Physics measurement principle for such an NDIR gas sensor which is supposed to always remain invariant. Wong's novel re-calibration invention (Wong, 2011d) extends his earlier disclosed Absorption Biased methodology (Wong, 2011b) for designing NDIR gas sensors by first transforming Graph 20 of Figure 15 for CO_2 into a curve that expresses the amount of target gas present in the sample chamber, P (ppm), as an nth order polynomial of parameter R where R is the normalized ratio of the Signal channel detector output over the Reference channel detector output.

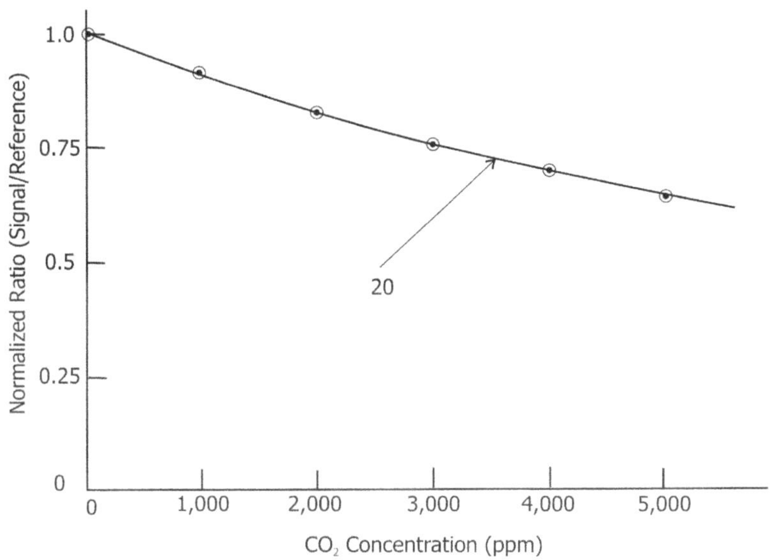

Figure 15. Normalization of de facto Calibration Curve with "Gamma" for an Absorption Biased (AB) Designed NDIR Gas Sensor.

For a third order polynomial, which is plenty accurate for most applications, this calibration curve transformation can be quantitatively expressed in terms of P (ppm) and Gamma as follows:

$$P\ (\text{ppm}) = A_0 + A_1 \times R + A_2 \times R^2 + A_3 \times R^3 \qquad (2)$$

$$\text{Gamma} = V_{S0}/V_{R0}\ (\text{no target gas in sample chamber}) \qquad (3)$$

$$R = (V_S/V_R)\ /\ \text{Gamma}, \qquad (4)$$

where V_S and V_R are respectively the Signal and Reference channel detector outputs when there is CO_2 (target gas) in the sample chamber. Note that in this transformation of the calibration curve for the sensor,

P (ppm) and Gamma of Equations (2) and (3) above represent, respectively, the invariant Physics principle portion and the inevitably variant components portion of the methodology. But since the parameter R is a function of Gamma (see Equation [4]), when there is a change in the value of Gamma in the course of time and not corrected, R will be affected and the calibration curve for the sensor accordingly will change leading to sensor output drifts. However, if for whatever reason the change in Gamma in the course of time is known, the value of R can be corrected and restored to its earlier value thereby preserving the original calibration curve for the sensor as represented by Equation (2). Under this circumstance, no output drifts will be detected from the sensor and it will stay accurate in the course of time.

Wong's novel re-calibration invention (Wong, 2011d) extends an additional step to his disclosed Absorption Biased methodology for designing NDIR gas sensors (Wong, 2011b) in order to achieve the objective of attaining a simple, easy and inexpensive re-calibration technique for such sensors. This is first done by reversing the expression of P (ppm) as a third order polynomial of R as depicted in Equation (2) above into one where R is expressed as a third order polynomial of P (ppm) without changing the value of Gamma as shown below:

$$R = B_0 + B_1 \times P + B_2 \times P^2 + B_3 \times P^3\text{; Gamma unchanged} \qquad (5)$$

For all AB designed NDIR gas sensors that can be re-calibrated with this novel re-calibration technique, they must carry both polynomials, namely Equations (2) and (5) in the memory of their Central Processing Unit (CPU).

Assume now that an NDIR gas sensor, in the current example CO_2, is calibrated with a calibration curve characterized by a third order polynomial with coefficients (A_0, A_1, A_2, A_3) and Gamma = G_0 (see Equations [2] and [3] above). As time goes by we recognize that the sensor no longer accurately detects CO_2 and we wish to re-calibrate this sensor to its original accuracy or calibration curve. To do this, one has to first prepare a secondary gas standard (in the current example CO_2) in the form an NDIR gas sensor which accurately detects and measures CO_2 concentration levels. This secondary gas standard and the sensor to be re-calibrated are then put together in the same still ambience (no winds or air movements within the space), preferably within a space volume of less than 1,000 cu. ft. The objective here is to make sure that both the secondary gas standard and the sensor to be re-calibrated sense or detect the same gas concentration level within this still space. The gas concentration level in the space as measured by the secondary standard is now transmitted (via wired or wireless communication) to the sensor to be re-calibrated. Upon receipt of this information, the sensor to be re-calibrated compares this received CO_2 concentration level with the one that it meanwhile also measures by determining the value of R and using the stored calibration curve (Equation [2]) and the Gamma value of G_0. If the gas concentration levels are found to be within the expected accuracy limit (say ± 50 ppm), a signal will be sent back to the secondary gas standard conveying the message that its measurement is good and the unit stays accurately calibrated. However, if the compared concentration levels lie outside the expected accuracy limit, then the unit needs re-calibration and will attempt to re-calibrate itself automatically as outlined below.

Using the received gas concentration level, it first attempts to calculate the corresponding R value using the stored reverse calibration curve (Equation [5]), namely (B_1, B_2, B_3, B_4). Using this newly calculated R value, the stored Gamma value of G_0 and the R value from its own gas measurement earlier that reports the inaccurate gas reading, a new Gamma, G_N, can now be determined. By simply replacing the old Gamma = G_0 with the newly determined G_N but retaining the original calibration curve (Equation [2]), namely (A_0, A_1, A_2, A_3), the sensor has just automatically re-calibrated itself.

Several factors are paramount and must be paid very close attention in order to make sure that Wong's re-calibration methodology works properly and is indeed valid. The first and most important factor is to make sure that the secondary gas standard in the form of another NDIR Absorption Biased gas sensor used in the re-calibration methodology is itself accurate. Such a secondary gas standard should be and could be frequently re-calibrated with a primary gas standard supplied by calibrated gas suppliers in the form of certified gas cylinders. Frequently re-calibrating the secondary gas standard with a primary one prior to its use can be easily achieved using the same re-calibrating technique described here. But since the secondary gas standard is designed according to the NDIR Absorption Biased methodology, it should remain stable most of the time and its re-calibration routine using the primary standard should be no more than a simple and easy accuracy check routine.

The second important factor to be paid close attention to is make sure that the gas concentration level sensed or detected by both the secondary gas standard and the sensor to be re-calibrated are exactly the same within specified measurement accuracy. In order for this to happen, the air space wherein the two sensors reside for the re-calibration routine must be still without air movements of any kind within it during the re-calibration time span. Another important precaution to be taken is to make sure that there is no target gas source present (e.g. CO_2 gas coming from a technician's breaths when re-calibrating CO_2 sensors) during the re-calibration routine.

A final important factor is that electrical communication (wired or wireless) between the secondary gas standard and the sensor to be re-calibrated must be robust and free from any interferences or disturbances. The sensor to be re-calibrated must receive an accurate gas concentration level of the relevant space from the secondary gas standard before it can properly re-calibrate itself. Once the sensor receives this information, by means of wireless or wired connection, it can automatically institute a self-recalibration routine. This self-recalibration can be automatic or only proceed if the difference between the gas concentration level measured by the sensor and the information received from the secondary gas standard (e.g., another NDIR gas sensor) exceeds a preselected threshold (e.g., the accuracy by which the secondary gas standard can detect the gas concentration of interest). Once the self-recalibration routine has started, the sensor can indicate whether any re-calibration has actually occurred and the result of such a recalibration. Information about the recalibration taken place can be transmitted by wired or wireless communication to a control point (e.g., a computer monitoring multiple sensors), stored in memory (where it may be obtained by later interrogation) or displayed visually, depending upon the designer's choice and any specialized needs by a user of the sensor.

The most significant advantage of Wong's invented re-calibration technique for NDIR gas sensors (Wong, 2011d) is the fact that it can re-calibrate a sensor in a particular space or location without the need to use any certified gas standards such as a 100 % dry nitrogen bottle or a 1,000 ppm CO_2 balanced with nitrogen gas cylinder. Instead, it simply uses the gas concentration level of the space or location wherein the sensor to be re-calibrated resides. In addition to tremendous labor savings because of the ease and simplicity of the currently advanced re-calibration technique, this also translates into very significant material savings (no standard gases needed) when commissioning buildings or re-calibrating installed gas sensors in commercial buildings.

5.5. Towards Perfection: Self-Commissioning NDIR Gas Sensors

From the discussion above one can see that we have come a long way in the understanding and exploitation of the NDIR gas measurement technique since its conception back in the 1950s. After over six decades of time have gone by, it is finally possible today to design NDIR gas sensors that have no significant output drifts over time using the AB methodology. Furthermore, one now also has a truly novel technique to simply, quickly and easily re-calibrate such sensors back to their original performance accuracy at any time and place without even the need for standard gases. These feats for the AB designed NDIR gas sensors are possible only because the Physics behind this gas measurement technology is fully understood. No other types of gas sensors that we know of today irrespective of from what technology or concept they come from can even come close to matching what the AB designed NDIR gas sensors can achieve in the world of gas measurement. However, knowing that nothing can be perfect but as long as there are mountains to climb and hurdles to overcome, there will be scientists and engineers out there trying hard to continue to strive for perfection. For the development of NDIR gas measurement technology there is no exception.

When Wong invented the novel recalibration technique for simply, quickly and easily restoring the measurement accuracy for AB designed NDIR gas sensors (Wong, 2011d), one of the very important precautions is to make sure that the target gas concentration level sensed by the sensor to be recalibrated is the same as that sensed by the calibration master which provides the ultimate standard gas level for the recalibration. To make sure that this condition could always be met, Wong in 2012 (Wong, 2012b) advanced the use of an air wand for drawing a sample of air very close to the sensor to be recalibrated to the sample chamber of the sensor inside the calibration master for measurement during the recalibration routine. The air wand is housed completely inside the calibration master and comprises a small air suction pump the inlet of which is connected to a small-bore telescopic tubing extendable to a length of six feet. Since the air sample drawn from the immediate neighborhood of the sensor to be recalibrated is delivered right to the sample chamber of the sensor inside the calibration master, any potential sources of the target gas even in the immediate neighborhood of the calibration master will not affect the measurement.

In 2011 Wong (Wong, 2011a) advanced a modified version of the Absorption Biased design methodology for NDIR gas sensors aimed at creating the same physical sample chamber for both the Signal channel and the Reference thereby eliminating any possible imbalances between them in the course of time. With both channels sharing almost the same physical sample chamber, any imbalances caused by the changing of sensor components characteristics over time as described earlier (see Section 5.4 of this Chapter above) will be reduced to an absolute minimum without causing any significant sensor output drifts. Wong's proposed new methodology utilizes like before the same narrow band-pass spectral filter for the detection of the gas of interest for both the Signal and the Reference channels. By so doing, the two channels always receive radiation of the same spectral content from the infrared source of the sensor convoluted with that from any external elements exposed to the sensor. While the same sample chamber through which the gas of interest to be detected flows is shared by the two channels, the detector package for the Reference channel is hermetically sealed with 100 % of the gas to be detected instead of 100 % N_2 as for the Signal detector. In so doing, the Reference channel is rendered almost completely "blind" to the presence or absence of the gas of interest flowing in the common sample chamber thereby creating an absorption bias between the two channels enabling the concentration of the gas of interest to be detected by ratioing the outputs of the two channels via calibration.

Taking another bold step towards perfection, Wong in 2012 (Wong, 2012c) invented the self-commissioning NDIR gas sensor. This invention is an extension of his earlier invention of the Absorption Biased (AB) design methodology (Wong, 2011b). In this invention two detectors of the same kind, each having an identical neutral band-pass filter to the target gas, are installed next to the Signal channel and Reference channel detectors as pairs in an AB designed NDIR gas sensor configuration. The two detectors are called the Standard Signal channel detector and the Standard Reference channel detector. Like Gamma of an AB designed NDIR gas sensor, the "Standard" Gamma is independent of the measurement Physics of NDIR gas sensors and is dependent only upon the performance characteristics of the sensor components. However, unlike Gamma, the "Standard" Gamma is also independent of the presence of any amount of target gas in the sample chamber. Consequently, "Standard" Gamma can be used to proportionally correct and update Gamma of the sensor as its components age over time thereby rendering such an AB designed

NDIR gas sensor self-commissioning or staying accurate over time after its initial calibration.

5.6. Looking into the Crystal Ball

Before we look into the crystal ball to find out what could possibly lie ahead in the world of NDIR gas measurement, we must first define what is "perfection" for this technology. We knew early on that while the NDIR gas measurement technique is one of the best available amid dozens of others it has quite a number of shortcomings. First, the absorption strength of the infrared signatures for gases detectable using this technique varies a great deal from one another. For example, while CO_2 has one of the strongest absorption strength as measured by its Absorbance value of 0.4, other important gases, such as CO, O_3, NO along with many indoor air polluting gases such as Formaldehyde, Benzene, Acrolein, Acetaldehyde etc., have absorption strengths order of magnitude smaller (absorbance < 0.01). To make things worse, the detection sensitivity required for some of these much weaker absorbing gasses is much higher than that for those gasses with medium to strong absorption signatures. Thus, at least in principle, to detect CO_2 gas with a required detection sensitivity of ± 25 ppm at a concentration level of 0 – 5,000 ppm is easy and straightforward. However, to detect CO gas with a required sensitivity of ± 1 ppm at a concentration level of 0 – 30 ppm is exceedingly difficult. To detect Ozone (O_3) with a required sensitivity of ± 0.05 ppm at a concentration level of 0 – 0.5 ppm is close to impossible. Since the technical foundation of NDIR gas measurement is based upon the Beer-Lambert law of absorption (see Equation [1] of Chapter 1), the only remedies to compensate for this situation are 1) use of a much longer path length sample chamber; 2) use of much higher sensitivity infrared detectors and 3) use of a much better infrared source with higher source temperature, larger emitting area and higher emissivity. But the use of one or more of these remedies will impose a lot of constraints on the size, performance and unit cost for the resultant NDIR gas sensors.

Second, the absorption strength of the infrared signatures for gases detectable using NDIR technique not only varies a great deal as discussed above, the locations of these signatures are also quite scattered throughout the 2 - 20 μ spectral region. That puts a lot of constraints on what types of infrared sources that can be used for what gases and the same applies to what types of detectors that would work for the detection of a particular gas. For example, the miniature incandescent source (light bulb) is small, bright, reliable, long-lived and inexpensive and has been used for decades as the most favored source for NDIR CO_2 sensors. However many gases that have their infrared signatures located above ~4.5 μ, like CO, NO, NO_2, SO_2, O_3 etc., to name just a few, cannot use this source because of its envelop that has very low spectral transmittance beyond 4.0 μ. As for detectors, although thermal detectors like pyroelectric and thermopile have "white" spectral response and can be used throughout the 2 - 20 μ spectral region for gas detection, their detection sensitivity as measured by their D^* is only adequate for gases with strong absorption signatures. For those gases with much weaker absorption signatures, one has to resort to using other types of much more expensive detectors (see Table 1 of Chapter 5), each with their own spectral response cutoffs extending beyond the infrared signature of the gases to be detected.

Finally, although the technical foundation of NDIR gas measurement is based upon the Beer-Lambert law of absorption, what the sensor is actually measuring is the density of target gas molecules between the source and the detector. The more target gas molecules are packed inside the column of space between the source and the detector, in other words the higher the density, the more relevant radiation from the source will be absorbed leading to a lower detector output. Since for a fixed volume of space, the density of the gas is both a function of its temperature and pressure. The output of the sensor therefore is dependent upon both the temperature and pressure of the environment wherein it is located. For some designs of NDIR gas sensors, like the AB designed methodology, the temperature dependence is negligible or could be easily compensated. For others, it could be as much as ± 1 % of reading per °C referenced to 25 °C. However, no sensor design can escape the pressure dependence. It is typically 0.18 % of reading per mmHg pressure above sea level. For some sensor designs like the AB designed methodology, this pressure dependence of the sensor output could be optionally and automatically compensated for with the use of a built-in pressure transducer chip.

From the discussion above, the pursuit of perfection for the NDIR gas measurement technology is meant nothing more than the continued exploitation of the technology by circumventing its weaknesses as described above. By taking full advantage of the fact that technologies of all kinds, especially in the world of micro and integrated electronics, nano-technology, intelligent software and faster processors etc., always go forward and push ahead, it will not be difficult for scientists and researchers in this field to circumvent some of the weaknesses of this gas measurement technology for bringing forth even better sensors in the days ahead.

Looking briefly into the crystal ball, one can already see that the pursuit in perfection for NDIR gas measurement technology is continuing. In 2011 Wong (Wong, 2011f) invented an intrinsically safe NDIR gas sensor in a can. This effort represents an important step towards the miniaturization of NDIR AB designed gas sensors aimed at special applications. In this invention an AB designed NDIR gas sensor is housed within a mechanical housing made up of a TO-5 can and a header with a diameter of ~1 cm. and ~0.8 cm high. This special AB NDIR gas sensor is for detecting reliably either Methane (CH_4) or water vapor (H_2O) thresholds inside coal mines for providing early warnings of accidental explosions or flooding in tunnels underground. These sensors have to be small, rugged, reliable and accurate with output stability over time, low power consumption, low unit cost and last but not least intrinsically safe.

Another important device in the pursuit of perfection for NDIR gas measurement technology that is forthcoming is an improved fire detector called "Mcfa" (Multi-criteria fire alarm) by detecting CO and CO_2 gases along with smoke at fire initiation and thereafter thereby potentially eliminating all false alarms, providing faster speed of response, protection from deadly CO inside households and enunciating automatically functional failures (Wong, 2011g). It has been mentioned before that the detection of CO with the kind of sensitivity required to perform as a fire sensor (± 1 ppm) using the NDIR AB designed methodology is indeed a formidable challenge. However, looking back on this technology for the past six decades and witnessing the journey it has gone through to what has become of it today, the sky is the limit !

One can readily imagine a future point in time, years from today, when NDIR gas sensors are found in widespread use, throughout the HVAC&R and fire detection industries, and in which many specialty NDIR gas sensors are used in many different applications to detect different gasses. Such sensors will be rugged, low-cost, accurate and inexpensive by today's standards. They either will not need re-calibration or they will be capable of easy recalibration, or even self-recalibration. They will be capable of being easily linked into data networks so their performance can be monitored and their measurements can be used to improve energy efficiency, save lives and improve the human condition. In fact, it is easy to imagine a day when the design of NDIR gas sensors will progress to the point that persons working in the field will wonder why it took so long for the industry to evolve. It did not happen overnight. We hope this book helps that day come more quickly.

References

Aoki et al. (1987), Gas analyzer with means for varying angle of incidence of light on interference filter, *US Patent* No. 4,662,755, May 5, 1987.

Avogradro, A. (1810), *Journal de Physique,* 73, 1810, pp. 58 – 76.

Bell, A. G. (1880), Upon the production of sound by radiant energy, *AM J Sci.,* 20, 1880, pp. 305-324.

Blau, H. H. Jr. (1974), Gas analyzer, *US Patent* No. 3,811,776, May 21, 1974.

Burch et al. (1974), Dual cell non-dispersive gas analyzer, *US Patent* No. 3,793,525, February 19, 1974.

Burough, I. G. et al. (1987), Method and apparatus for detecting gas, *US Patent* No. 4,709,150, November 24, 1987.

Eden, D. D. (1981), Optical modulation with vanadium oxide thin films, *US Patent* No. 4,283,113, August 11, 1981.

French, J. M. (1974), Gas analyzer, *US Patent* No. 3,811,776, May 21, 1974.

Hayes, T. J., Merrick, E. B. (1976), Continuous, non-invasive measurements of blood oxygen levels, *Hewlett Packard J.,* 28, 1976, pp. 2-10.

Infrared Analysis, Inc. (1995), Infrared Spectra for Quantitative Analysis of Gases, *Anaheim*, CA 92801, 1995.

Jaffe, B. M. (2008), Infrared measurement of carbon dioxide in the human breath: breathe-through devices from Tyndall to the present day, Technology, *Computing and Simulation*, Vol. 107 No. 3, 2008, pp. 890-904.

Lampert, et al. (2001), Dynamic infrared sensor for automotive pre-vaporized fueling control, *US Patent* No. 6,237,575, May 29, 2001.

Leslie, J. (1804), Leslie's cube (named after Sir John Leslie who experimented with radiant heat in 1804), 1804.

Luft, K. F. (1943), 50 years NDIR gas analysis, Tm Technisches Messen, 60, 1943, pp. 363-371.

Maiden, A. J. (1985), Non-dispersive optical determination of gas concentration, *US Patent* No. 4,500,207, February 19, 1985.

Pembrook et al. (1975), Electronically compensated rotating gas cell analyzer, *US Patent,* No. 3,878,107, April 15, 1975.

Pfund, A. H., Gemmill, C. L. (1940a), An infrared absorption method for the quantitative analysis of respiratory and other gases, *Bull Johns Hopkins Hosp*, 67, 1940, pp. 61-65.

Schoennauer et al. (1989), Resonant piezoelectric chopper for infrared radiation, *US Patent* No. 4,850,697, July 25, 1989.

Sternberg et al. (1974), Method and system for the infrared analysis of gases, *US Patent* No. 3,790,798, February 5, 1974.

Thatcher, J. B. (1988), Apnea Monitor, *US Patent* No. 4,738,266, April 19, 1988.

Tse, C. W. et al. (2007), Low-cost noise-immune wireless methodology for demand control ventilation (DCV) applications, *US Patent* No. 7,239,625, July 3, 2007.

Tyndall, J. (1859), On the transmission of heat of different qualities through gases of different kinds, *Proc. R Inst GB,* 3, 1859, pp. 155-158.

Tyndall, J (1865), On radiation, The "Rede" lecture delivered in the Senate-House before the University of Cambridge on Tuesday, May 16, 1865, *Longman, Green, Longman, Roberts & Green,* London, 62 p.

Warmack et al. (2009), private communication.

Wong, J. (1986), Self-calibrating carbon dioxide analyzer, *US Patent* No. 4,578,762, March 25, 1986.

Wong, J. (1987), Non-dispersive gas analyzer having no moving parts, *US Patent* No. 4,694,173, September 15, 1987.

Wong, J. (1988a), Oxygen measurement using visible radiation, *US Patent* No. 4,730,112, March 8, 1988.

Wong J. (1988b), Compact apparatus for measuring absorption by a gas, *US Patent* No. 4,756,622, July 12, 1988.

Wong, J. (1989), High speed infrared shutter, *US Patent* No. 4,795,240, January 3, 1989.

Wong, J. (1991a), Spectral ratioing technique for NDIR gas analysis using a differential temperature source, *US Patent* No. 5,026,992, June 25, 1991.

Wong, J. (1991b), Simple fire detector, *US Patent* No. 5,053,754, October 1, 1991.

Wong, J. (1991c), Gas sample chamber, *US Patent* No. 5,060,508, October 29, 1991.

Wong, J. (1992a), Rapid fire detector, *US Patent* No. 5,103,096, April 7, 1992.

Wong, J. (1992b), Gas sample chamber, *US Patent* No. 5,163,332, November 17, 1992.

Wong, J. (1993), Multi-channel gas sample gas chamber, *US Patent* No. 5,222,389, June 29, 1993.

Wong, J. (1994a), Functional testing method for toxic gas sensors, *US Patent* No. 5,335,534, August 9, 1994.

Wong J. (1994b), Diffusion-type gas sample chamber, *US Patent* No. 5,340,986, August 23, 1994.

Wong, J. (1994c), NDIR gas analysis using spectral ratioing technique, *US Patent* No. 5,341,214, August 23, 1994.

Wong, J. (1994d), Self-calibration of an NDIR gas sensor, *US Patent* No. 5,347,474, September 13, 1994.

Wong, J. (1995a), Gas sample chamber for use with a source of coherent radiation, *US Patent* No. 5,384,640, January, 24, 1995.

Wong, J. (1995b), Laser assisted quasi-blackbody radiation source, *US Patent* No. 5,408,101, April 18, 1995.

Wong, J. (1995c), Enhanced path length gas sample chamber, *US Patent* No. 5,414,264, May 9, 1995.

Wong, J. (1995d), NDIR gas sensor, *US Patent* No. 5,444,249, August 22, 1995.

Wong, J. (1995e), Enhanced pathlength gas sample chamber, *US Patent* No. 5,453,621, September 26, 1995.

Wong, J. (1996), Diffusion-type gas sample chamber, *US Patent* No. 5,502,308, March 26, 1996.

Wong, J. (1997a), False alarm resistant fire detector with improved performance, *US Patent* No. 5,592,147, January 7, 1997.

Wong, J. (1997b), Passive infrared analysis gas sensor, *US Patent* No. 5,650,624, July 22, 1997.

Wong, J. (1997c), Practical and improved fire detector, *US Patent* No. 5,691,704, November 25, 1997.

Wong, J. (1998), Passive and active infrared analysis gas sensors and applicable multichannel detector assemblies, *US Patent* No. 5,721,430, February 24, 1998.

Wong, J. (2007a), Passive NDIR gas sensor fire detector, *US Patent* No. 7,176,460, February 13, 2007.

Wong, J. (2007b), Passive NDIR carbon dioxide sensor fire detector, *US Patent* No. 7,186,979, March 6, 2007.

Wong, J. (2007c), Ultra low power NDIR carbon dioxide sensor fire detector, *US Patent* No. 7,214,939, May 8, 2007.

Wong, J. (2007d), Method for detecting a gas species using a super tube waveguide, *US Patent* No. 7,259,374, August 21, 2007.

Wong, J. (2008a), Simple multi-channel NDIR gas sensors, *US Patent* No. 7,329, 870, February 12, 2008.

Wong, J. (2008b), Ultra low cost NDIR gas sensors, *US Patent* No. 7,358,489, April 15, 2008.

Wong, J. (2011a), Saturation filtering NDIR gas sensing methodology, *US Patent* No. 8,003,944, August 23, 2011.

Wong, J. (2011b), Absorption biased NDIR gas sensing methodology, *US Patent* No. 8,143,581, March 27, 2012.

Wong et al. (2011c), Zero drift NDIR gas sensors, *Sensor Review,* Vol. 31, No. 1, 2011, pp. 70-77.

Wong, J. (2011d), Re-calibration methodology for NDIR gas sensors, *US Patent* No. 8,178,832, May 15, 2012.

Wong, J. (2011e), Calibration methodology for NDIR dew point sensors, *US Patent* No. 8,148,691, April 3, 2012.

Wong, J. (2011f), An intrinsically safe NDIR gas sensor in a can, *US Patent* No. 8,003,945, August 23, 2011.

Wong, J. (2011g), False alarm resistant and fast responding fire detector, *US Patent* No. 8,077,046, December 13, 2011.

Wong, J. et al. (2012a), Re-calibration methodology for NDIR Gas Sensors, accepted for publication in *Sensor Review,* Vol. 32-3, 2012.

Wong, J. (2012b), Air sampler for recalibration of absorption biased designed NDIR gas sensors, *Attorney Docket* No. RLA35. 167, Notice of Allowance to issue received June 4, 2012.

Wong, J. (2012c), Self-commissioning NDIR gas sensors, *Attorney Docket* No. RLA35. 168, Notice of Allowance to issue received May 31, 2012.

Yamada, Y. (1986), Gas analyzer with compact cell structure, *US Patent* No. 4,605,855, August 12, 1986.

Index

G

H

I

L

M

N

O

www.ingramcontent.com/pod-product-compliance
Ingram Content Group UK Ltd.
Pitfield, Milton Keynes, MK11 3LW, UK
UKHW020207290926
15470UKWH00001B/54

9 788461 597321